INSIDE CATIA

Paul Carman
and Paul Tigwell

INSIDE CATIA ®

By Paul Carman and Paul Tigwell

Published by:
OnWord Press
2530 Camino Entrada
Santa Fe, NM 87595-4835 USA

Carol Leyba, Publisher
David Talbott, Acquisitions Editor
Barbara Kohl, Associate Editor
Cynthia Welch, Production Manager
Michelle Mann, Production Editor
Liz Bennie, Director of Marketing
Lauri Hogan, Marketing Manager
Lynne Egensteiner, Cover designer, Illustrator

All rights reserved. No part of this book may be reproduced or transmitted in any form or by any means, electronic or mechanical, including photocopying, recording, or by any information storage and retrieval system without written permission from the publisher, except for the inclusion of brief quotations in a review.

Copyright © Paul Carman and Paul Tigwell

First Edition, 1998

SAN 694-0269

10 9 8 7 6 5 4 3 2 1

Printed in the United States of America

Library of Congress Cataloging-in-Publication Data

Carman, Paul, 1954-
    Inside CATIA / Paul Carman and Paul Tigwell.
       p. cm.
    Includes index.
    ISBN 1-56690-153-7
    1. Computer-aided engineering—Software. 2. CATIA (Computer file).
  I. Tigwell, Paul, 1953-   . II. Title.
  TA345.C37   1998
  670'.285'5369—dc21

                                                            97-47325
                                                               CIP

Trademarks

CATIA is a registered trademark of Dassault Systemes. OnWord Press is a registered trademark of High Mountain Press, Inc. Other products are mentioned in this book that are either trademarks or registered trademarks of their respective corporations. OnWord Press and the authors make no claim to these marks.

Warning and Disclaimer

This book is designed to provide information about CATIA. Every effort has been made to make the book as complete, accurate, and up to date as possible; however, no warranty or fitness is implied. The information is provided on an "as is" basis. The authors and OnWord Press shall have neither liability nor responsibility to any person or entity with respect to any loss or damages in connection with or arising from the information contained in this book.

About the Authors

Paul Carman, a designer with 20,000 hours of experience working with CATIA, obtained a higher national certificate in mechanical engineering from Reading College of Technology, and is a member of the Institute of Incorporated Mechanical Engineers. He joined the Joint European Torus (JET) nuclear fusion experiment at Culham (near Oxford, England) in 1976, and has been providing CATIA training since January 1987. At present, Paul serves as co-coordinator of CATIA training and problem solving at JET.

After obtaining a higher national certificate in mechanical engineering from Reading College Technology, **Paul Tigwell** worked in the drawing offices of various companies before joining JET in 1987. Paul has served as co-coordinator of CATIA training and problem solving at JET since 1995.

Acknowledgments

I wish to express my appreciation to the JET Project for choosing CATIA and thus giving us the opportunity to work with and provide training in CATIA; to

Dr. Michael Pick at JET for the introduction to OnWord Press; to Henri Duquenoy (head of the JET Design Office) for his cooperation; to Mark Claxton, Dave Robson, and particularly Krishan Purahoo for system support; and to our good friend Tim Potter for our original CATIA training.

Next, I would like to thank my family and friends for their support and encouragement during the writing of *INSIDE CATIA*, as well as my partner in this venture, Paul Tigwell ("Tiggers"), who helped to make it a pleasure. Finally, I wish to dedicate this book to my wife, Jacqueline ("Ginge").

Paul Carman

I would like to thank my wife, Sandra, and my boys, Dean and Alex, for their patience and encouragement during the writing of this book. Thanks must also go to my family and friends for their encouragement and also to the "other" Paul for being the ideal writing partner.

Paul Tigwell

Contents

Introduction *XIII*

CATIA Described **xiii**
Scope of INSIDE CATIA **xvi**
Book Organization **xvii**
How to Use INSIDE CATIA **xviii**
Typographical Conventions **xix**
Installing Models from Companion CD-ROM **xx**

Chapter 1   Setting up the CATIA Environment *1*

Basic Hardware **1**
Logging On **3**
Work Area **6**
    Palettes **7**
    Starting a New Model **8**
Graphical Representation of 2D Elements **10**

Chapter 2   Generating 2D Graphic Elements *11*

Alternatives for Creating Graphical Elements **11**
Using POINT, LINE, and CURVE2 **15**
    Setting Up **15**
    Model Manipulation **16**
Exercise 1: Creating a Rectangle **17**
Exercise 2: Adding Lines, Points, and Curves **20**
Exercise 3: Creating an Angle Section **22**
Exercise 4: Creating an I Beam **23**
Summary **24**

Chapter 3 Modifying 2D Graphic Elements 25

Alternatives for Modifying Graphical Elements   25
Exercise 1: Creating a Simple Box Using Unlimited
   Lines   30
Exercise 2: Exploring ERASE Function Options   31
Exercise 3: Creating an Angle Section with Radius
   Corner   32
Exercise 4: Creating an I Beam
   with Radius Corners   34
Exercise 5: Creating an I Beam with Dot-dashed Center
   Lines and Hidden Axis   36
Exercise 6: Creating a Universal Beam   40
Exercise 7: Creating a Side Plate Using Additional
   Options   42
     Creating Palettes   45
Summary   48

Chapter 4 Analyzing 2D Graphic Elements 49

Exercise 1: Introduction to Alphanumeric Window   51
Exercise 2: Relative Analysis   53
Exercise 3: Creating Palettes   54
Summary   56

Chapter 5 Managing 2D Graphic Elements 57

TRANSFOR in DRAW Mode   58
GROUP in DRAW Mode   59
DETAIL in DRAW Mode   61
LIBRARY in DRAW Mode   63
SETS in DRAW Mode   65
LAYER in DRAW Mode   66
IMAGE in DRAW and SPACE Modes   67
Exercise 1: Using TRANSFOR   69
Exercise 2: Using GROUP   73
Exercise 3: More Transformation Options   75
Exercise 4: Using the DETAIL Function   76
LIBRARY Function   80
Exercise 5: Using SETS   80

Exercise 6: Using LAYER **81**
Exercise 7: Using IMAGE **85**
Summary **88**

Chapter 6   Creating Multiple Views *89*

AUXVIEW **90**
COMBIVU **91**
DRAFT **92**
AXIS **93**
Exercise 1: Using AUXVIEW and COMBIVU **94**
   Creating the Front View **94**
   Side View **99**
   Finishing Details **106**
   Translating Views Around the Screen **108**
   Plan View **108**
   Isometric View **112**
Exercise 2: Using AXIS in DRAW Mode **116**
   Multi-selection Options **119**
Exercise 3: Using AXIS with ANALYSIS **120**
Exercise 4: Using DRAFT and AUXVIEW **122**
Summary **124**

Chapter 7   Annotating Drawings *125*

TEXTD2 **125**
DIMENS2 **133**
PATTERN **141**
DRWSTD (Draw Standard) **143**
Exercise 1: Using TEXTD2 and DIMENS2 **146**
   Creating a Drawing Sheet Blank **146**
   Creating a Title Block **148**
   Annotating a Title Block **150**
   Inserting a Drawing Blank Detail in the Master Workspace **154**
   Creating a Symbol of Drawing Blank **156**
   Dimensioning Views **158**
   Creating Leaders **162**
   Creating Notes **164**

Exercise 2: Using DRAFT and PATTERN **166**
Summary **172**

Chapter 8   **Hardcopy Output Using PLOT** *173*

PLOT **173**
Exercise 1: Creating a Full-size Quick Plot **176**
Exercise 2: Creating a Multiple Window Plot in Quick Mode **182**
Exercise 3: Creating and Modifying a Multiple Window Plot in FILE Mode **183**
Exercise 4: Producing a Screen Capture **186**
Summary **188**

Chapter 9   **Generating and Modifying SPACE Graphic Elements** *189*

POINT **190**
LINE **191**
CURVE2 **194**
ERASE **197**
LIMIT1 **198**
GRAPHIC **199**
PLANE **200**
Generating and Modifying SPACE Graphic Elements **202**
  Setup for Exercises **202**
  3D and 2D SPACE **202**
Exercise 1: Creating I Beam Geometry in SPACE Mode **204**
Exercise 2: Creating Side Plate Geometry in SPACE Mode **207**
Exercise 3: Using POINT and GRAPHIC **208**
Exercise 4: Creating a Wireframe Model **210**
Summary **213**

Chapter 10   **Analyzing 3D Graphic Elements** *215*

Exercise 1: Using the ANALYSIS Function **216**
Summary **220**

Chapter 11   Creating Solids *221*

    Background on Solids   **221**
        Solid Representations   **221**
        Mock-up and Exact Solids   **221**
    SOLIDM   **222**
    SOLIDE   **225**
    Exercise 1: Creating a Complex Solid
      Using Primitives   **227**
        Solid Base   **227**
        Conical Solid   **229**
        Cylinder   **230**
        Sphere   **231**
        Drafting Base Solid Sides   **231**
        Joining Separate Primitives to Form Single Solid   **232**
        Fillet Radius Around Top Edge of Base   **232**
        Performing Local Transformations of the 3D
           Model   **233**
        Create Shell   **235**
        Cutting Shell Solid in Half   **236**
        Modifying Solid Parameters   **237**
    Exercise 2: Creating a Solid I Beam   **238**
    Exercise 3: Creating the Solid Side Plate   **240**
    Exercise 4: Creating a Solid Wheel   **244**
    Exercise 5: Creating a Solid Nut   **246**
    Summary   **249**

Chapter 12   Analyzing Solids *251*

    Exercise 1: Analyzing Solids
      Using the ANALYSIS Function   **251**
    Exercise 2: Analyzing Solids Using
      the SOLIDE Function's ANALYSIS Option   **253**
    Exercise 3: Analyzing the Position
      of Several Solids   **257**
    Exercise 4: Combining Analysis Results   **260**
    Summary   **261**

Chapter 13 Managing Solids *263*

DETAIL in SPACE Mode **265**
MERGE in SPACE Mode **266**
Exercise 1: Duplicating and Positioning the Side Plates **268**
Exercise 2: Duplicating and Positioning the Wheels **269**
Exercise 3: Merging the Spindle Assembly Parts into the Wheel Model **271**
Exercise 4: Using IMAGE to Create Windows and Screens **274**
Summary **276**

Chapter 14 2D/3D Integration *277*

Using DRW➔SPC, SPC➔DRW, and SPC➔DR2 **277**
Exercise 1: Transferring Elements from DRAW to SPACE **280**
Exercise 2: Creating Solids from DRAW Geometry **283**
Exercise 3: Creating DRAW Geometry by Transferring from SPACE to DRAW **285**
Exercise 4: Producing DRAW Views from Solids **287**
Summary **290**

Chapter 15 Working in a Multi-Model Environment *291*

Exercise 1: Creating a Multi-model Environment **293**
Exercise 2: Copying Elements Between Models **298**
Exercise 3: Using BREAKOUT **298**
Exercise 4: Creating Work Areas **300**
Summary **302**

Chapter 16 Creating Orthographic Views from Solid Models *303*

AUXVIEW2 **303**
TEXT **309**
AUXVIEW **310**
Exercise 1: Using AUXVIEW **310**
Exercise 2: Using AUXVIEW2 **314**

Exercise 3: Updating Views **325**
Exercise 4: Modifying Views **326**
  Filtering Solids in a View **326**
  Modifying the Visualization of Solids **329**
  Clipping a View **330**
  Breaking Out Solids from within a View **331**
Summary **333**

Chapter 17   Parametric Modeling 335

Exercise 1: Creating a Parameterized Profile
  of Beam Trolley I Beam **337**
Exercise 2: Extracting DRAW Views
  with Autodimensioning **341**
Exercise 3: Using Parametric Dimensions
  to Position Features **343**
Note on Dynamic Sketcher **347**
Summary **347**

Chapter 18   Housekeeping Tips 349

Exercise 1: Using the ERASE Function **352**
  Erasing Unnecessary Details or Symbols **353**
Exercise 2: Exploring the /CLN Command **354**
Exercise 3: Exploring the KEEP Function **356**
Exercise 4: Using the IDENTIFY Function **357**
Summary **358**

Appendix A   Element Identifiers Used in CATIA 359

Appendix B   Keywords Used for Multiple Selection 363

Appendix C   CATIA General Commands 369

Appendix D   Default Function Palette 373

Appendix E   CATIA Pull-down Menus 375

File **375**

Select **376**
View **377**
Filter **379**
Options **379**
Tools **380**
Window **381**
Help **382**

Appendix F   CATIA Utilities *385*

Appendix G   Engineering Symbols *390*

INDEX   *391*

Introduction

CATIA Described

CATIA is produced by Dassault Systemes in Paris, and is considered to be among the most powerful CAD systems in use today. CATIA, a highly interactive 3D CAD/CAM system, is an acronym for computer aided three-dimensional integrated application. The software provides two-dimensional drafting and three-dimensional modeling facilities together with several design analysis tools and comprehensive manufacturing facilities.

Any CAD system is only as good as the person who uses it. With this in mind, the authors have attempted to produce a book for self-instruction as well as an intermediate CATIA solutions setup guide. This book provides the tools for a user to produce anything from 2D drawings through complex solid models. *INSIDE CATIA*, along with the *CATIA Reference Guide*, provide reference works thus far unavailable.

The authors have used CATIA for ten years. In their opinion, the software is without rival. The JET Project at Culham, England, is testimony to CATIA's versatility. JET is the most successful nuclear fusion experimental machine in the world; the present configuration would not have been possible without CATIA. CATIA is also used extensively by some of the world's leading aerospace and automotive manufacturers.

Inside the torus (nuclear fusion experiment, Culham, England).

CATIA Described

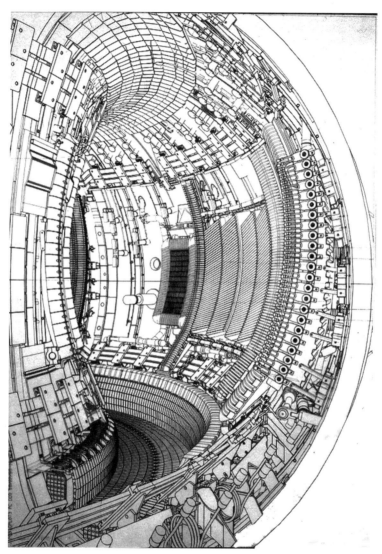

Cut-away view of a portion of the torus designed in CATIA.

Scope of *INSIDE CATIA*

This book can serve as a tutorial for designers with no CAD background to become competent CATIA users. In this context, "competent CATIA users" means the ability to produce anything from 2D drawings to complex 3D solid model assemblies.

Unlike some CAD systems, CATIA does not force you to approach a task in a particular way. CATIA is not strictly a production engineering widget design tool, although it is more than capable of fulfilling this role. It is a comprehensive design tool box, and in the right hands it is formidable.

After ten years of use we continue to be regularly impressed by the power of CATIA. CATIA is enjoyable to use, and effectively invites the user to explore and experiment. The user should always remember that there are many ways to achieve a particular result, and you can quickly become increasingly competent via experimentation. CATIA users should also keep in mind that there is not always a right or wrong way of achieving an end result. Instead, there may be many different ways of achieving a desired outcome, and the right way is defined as the way in which the user feels most comfortable.

CATIA's structure is such that the software can used for creating simple 2D drawings or a complete aircraft, automobile, or even a nuclear fusion research experiment. Given CATIA's diversity and the vast number of available CATIA modeling tools, explaining and providing examples of all tools in a moderately sized book is impossible. Therefore, the objective of this book is to provide a tutorial for the basic and intermediate level of available tools.

By demonstrating the power of selected tools and how they are used, the authors hope to encourage you to experiment with them to ultimately capture their full potential. In this fashion, users are then invited to explore CATIA's specialized tools. The emphasis of the book is to present diverse methods of achieving an objective, leaving the designer free to select the appropriate one for the project at hand. Once you have mastered the basic skills necessary to drive CATIA, you can develop expertise in one or more specialized areas.

Book Organization

In this book computer jargon is kept to an absolute minimum because the work is aimed at users rather than programmers. The book is organized into 18 chapters arranged sequentially from creating basic 2D drawings and setting up the work environment, to creating 3D models and manipulating a multi-model environment.

Chapter 1 covers the basics of switching CATIA on and off, saving your work, how to communicate with CATIA and vice versa. By the end of the chapter, the user should be able to navigate CATIA menus and command structure.

Chapter 2, "Generating Graphic Elements," demonstrates how to generate basic 2D draw elements such as lines, points, and curves. In Chapter 3, "Modifying Graphic Elements," elements generated in Chapter 2 are modified. Modifications include erasing elements, extending lines and curves, shortening lines and curves, and changing the graphical representation of elements on the screen and for printing (e.g., solid, dotted, chain dotted lines, among other options). This chapter also includes an introduction to multi-selection, the ability to select groups of elements.

Analysis of graphic elements is the topic of Chapter 4. Chapter 5 demonstrates moving geometry around the screen, hiding geometry, and managing geometry used more than once in a model or in other models.

Chapter 6, "Creating Multiple Views," provides an introduction to producing complex geometry across related views without the need for 3D geometry. Chapters 7 and 8, "Annotating Drawings" and "Hardcopy Output Using PLOT," respectively, demonstrate how to add text and dimensions to a drawing, and make a reproducible copy of your work.

Chapters 9, 10, and 11 provide demonstrations on creating and managing elements in the 3D space environment. These chapters cover steps similar to those discussed in Chapters 2, 3, 4, and 5, but in 3D rather than 2D workspace.

Chapters 12 and 13 are focused on solid modeling: how to create, analyze, and manage solid models. Chapter 14, "2D/3D Integration," is focused on creating solids from 2D geometry and creating 2D views from 3D solids and geometry. These facilities allow users to create finished 2D drawings of design work created in 3D, as well as create 3D models based on design work carried out in 2D.

Chapter 15 covers working in a multi-model environment, including simultaneous display of more than one model, using geometry in one model to create and manage geometry in another, and transferring geometry and solids from one model to another. Chapter 16 is focused on creating orthographic views from solid models.

Chapter 17 provides an introduction to parametric modeling, or the ability to drive 3D models from 2D geometry and vice versa. Chapter 18 provides numerous housekeeping tips useful in the CATIA environment.

The book's seven appendices include abbreviations used for element identification, keywords used in multiple selections, general commands that can be entered via the input information area, the default function palette, pull-down menus, utilities, and engineering symbols. The book closes with a complete topical index.

How to Use INSIDE CATIA

INSIDE CATIA was designed as a hands-on tutorial. You are encouraged to read the text and examine the example drawings, while also working through the exercises at a CATIA workstation, regardless of how simple or basic they may appear.

Exercises are kept separate from explanatory material. Thus, you can work through the exercises without reading all of the text. For example, if you wish to repeat an exercise to more thoroughly understand a particular function, you can do so without the necessity of rereading explanatory text.

Exercises

Every chapter contains one or more exercises. All are based on a target drawing and provide a step-by-step guide to accomplishing the designated task. Some exercises contain more than one series of steps, thereby illustrating multiple means for producing a given result.

As you proceed through the chapters, models are created and stored to be used again to produce assemblies. In almost all instances, the final results of exercises are stored on the companion CD. You can access a model (file) from the CD to compare results. In some instances, you may wish to commence an exercise with a model stored on the CD rather than backtrack and create it yourself.

On Measurement Units

The text and exercises throughout the book have been written based on the metric (mm) system of measurement, and using a scale of 1. The fact that the exercises are based on the metric system makes no difference to the functionality of CATIA, and the exercises would be completed in exactly the same way if they were based on customary U.S. measures (in., ft., etc.).

If your system is set up to work in inches, you can convert the units used throughout the book (i.e., 1" = 25.4 mm, 1mm = 0.0394"). This conversion calculation can be undertaken directly in the input information area (see Chapter 1). In other words, 38mm input into an inch-based system could be entered as 38/25.4 or 38*0.0394.

The other option available to you is to change the model units used on your system with the STANDARD > MODEL function, and then change the UNIT CONVERSION settings in the GEOMETRIC STANDARDS panel. Use this option only with the approval of your system administrator: once a model has been created you cannot change the model units. Therefore, if your system uses an automatic start-up model, you would be unable to change the working units.

> ✤ *NOTE*: *All online documentation available through FRAMEVIEWER (which should be available on your workstation) is written for metric units.*

Typographical Conventions

The names of CATIA commands and information appearing in the CATIA message area are all capitalized (e.g., AUXVIEW, COMBIVU, DRAFT). Interface items such as menu and palette names and terminology specific to CATIA are typically capitalized.

Command sequences, or functions followed by options, are usually separated by an angle bracket (>). In contrast, command sequences showing the selection of an option, followed by one or more suboptions, are separated by a pipe (|). Examples appear below.

FILE > READ

LAYER > FILTER | APPLY | DIRECT

PARAM3D > PROFILE | CONSTRN

User input and names for files in running text are italicized.

Read the *BEAM TROLLEY SOLID I BEAM* model saved in Chapter 11.

Key in *150/2.*

Keys on a standard keyboard mentioned in running text are enclosed in angle brackets. Examples follow.

<Enter>

<Ctrl>

<F4>

<r>

> ✥ **NOTE:** *Information on features and tasks that is not straightforward, immediately obvious, or intuitive appears in notes.*

> ✓ **TIP**: *Tips on command usage, shortcuts, and other information aimed at saving you time appear like this.*

> ✗ **WARNING**: *The warnings appearing in this book are intended to help you avoid committing yourself to results that you may not have intended.*

Installing Models from Companion CD-ROM

The companion CD contains a full set of exported CATIA models of exercises in the book. There are start and target models for each exercise. Access the

Installing Models from Companion CD-ROM

models for practice sessions or if you experience difficulties. The procedure for installing the files is described below.

You may need the assistance of your system administrator because many systems permit only the superuser (root) to mount CDs. For UNIX users, the CD should be made available as if it were a file system. In most cases, the following standard mount command can be used.

```
# mount options CD-ROM_device mount_point
```

The *CD-ROM_device* name varies depending on the type of system. If you do not know the device name, consult the documentation that shipped with your system. The *mount_point* is simply a directory that will become the parent directory of the CD when it is mounted.

1. Insert the CD into the local CD drive on your workstation.
2. Create a directory for the CD, if it does not already exist, as follows:

```
# mkdir /CDROM
```

Depending on the hardware platform and operating system you are using, mount the CD drive by issuing one of the commands appearing below.

For IBM (AIX 3.x/4.x):

```
# mount -rv cdrfs /dev/cd0 /CDROM
```

For SGI:

```
# mount -rt iso9600 /dev/scsi/scCdu10 /CDROM
```

where C is scsi controller number, and u is CD-ROM unit number.

The above values can be identified by entering the following command:

```
/bin/hinv
```

For Digital UNIX:

```
# mount -t cdfs /dev/rzua /CDROM
```

where u is CD-ROM unit.

For all other platforms consult your system administrator and/or your operating system documentation.

Once the CD is mounted, the models can be imported by taking these steps.

1. From the CATIA_V4 window on the desktop, double-click on the CatUtil icon.

2. In the Catutil window, double-click on CATIMP (Import CATIA models or libraries to a CATIA site). Change the settings to match the panel in the following illustration. Note the file name, *icatia.exp*, in the INPUT FILE area on the panel. The companion CD contains a single exported file. This file in turn contains 122 CATIA models of varying sizes.

CATIMP panel.

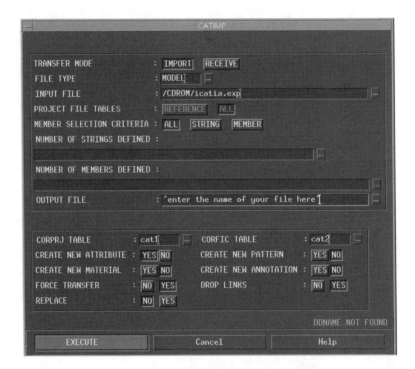

NOTE: *Set the OUTPUT FILE to the CATIA file in the location you wish for the exercise models.*

3. Click on the EXECUTE button, and then click on the OK button in the Execute window.

All models from the CD should have been imported to your selected file. If you have difficulties with any of the above steps, contact your system administrator.

1 Setting up the CATIA Environment

This chapter is focused on getting started in CATIA. Topics include the log-on procedure, work area layout, how selected tools are used, how selections are made, and how information is entered.

Basic Hardware

It is assumed that CATIA version 4, release 1.6 or later has been successfully installed on a suitable workstation by your system administrator, and that the three minimum basic hardware items listed below are available.

- Color graphics display (monitor or screen).

- PC101 type keyboard. (Other keyboard types can be configured for use with CATIA. See your system administrator for details.)

- Three-button mouse. Button 1 is used to select or pick; button 2, to indicate or point; and button 3, to drag an item previously selected with button 1.

Appearing below is a short list and description of additional hardware items that you or your organization may choose to use for working with CATIA.

- Four-button puck or mouse and tablet. The tablet mouse or puck is advantageous in that the CATIA screen area is fixed by the tablet so that the position of the cross hairs on the puck in relation to any point on the screen is absolute. The four buttons are used as follows: button 1 is used to select or pick; button 2, to indicate or point; button 3, to drag; and button 4, to switch on an alphanumeric window. (Window usage is described later in this book.)

➼ **NOTE:** *The next four illustrations are screen grabs of solid CATIA models of the listed hardware items.*

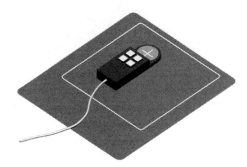

Four-button puck or mouse and tablet.

- Dials. The dials allow zooming, panning, rotations, and translations. Dials facilitate control of the work area and are strongly recommended.

Dials.

- Spaceball. The spaceball performs translations, rotations, and zoom/pan, and can be used with or instead of dials.

Spaceball.

- Lighted program function keyboard. The LPFK keyboard is comprised of a pad of 32 programmable keys which can be set to correspond to CATIA functions and instructions. The use of this hardware item enables much faster selection of commonly used functions and instructions and is strongly recommended.

LPFK (lighted program function) keyboard.

Logging On

The log-on procedure may vary slightly, depending on your particular setup. Only the selections necessary to get you started will be covered in this chapter. The many additional facilities available on the panels discussed in this section will be discussed later in the book. Take the following steps to log on and begin a CATIA session.

> **NOTE:** *The AIX 4.1.3 or 4.1.4 operating system is assumed. Next, throughout the book it is assumed that you are using a three-button mouse.*

1. To commence the log-on procedure, verify that the workstation and monitor are on. A window will display asking you to enter your user ID. Type the user ID and press the <Enter> key. The next window requests

4   **Chapter 1: Setting up the CATIA Environment**

your password. Input the password and press <Enter>. The next three illustrations show the AIX desktop and front panel, and two additional windows that display when the password is entered.

AIX desktop front panel.

CATIA Environment window.

CATIA V4 Options window.

2. In the CATIA Environment window the YOUR option icon is represented by a figure sitting in front of a workstation. If not already highlighted, click on the YOUR icon with mouse button 1.

Logging On

> ⇨ **NOTE:** *Henceforth, whenever the term "click on," "double-click," or "select" appears, use mouse button 1.*

3. To begin a CATIA session, double-click on the CatMotif icon in the CATIA window.

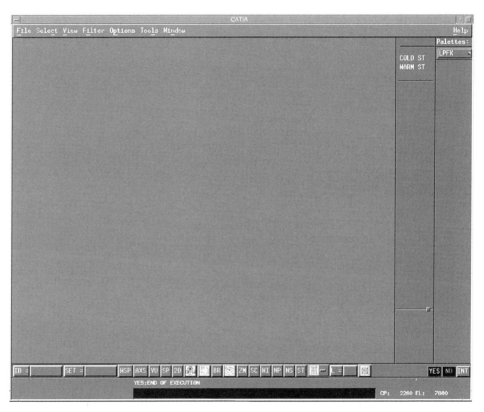

CATIA Start window.

4. The next step is to decide on a cold or warm start. A cold start takes you into an empty workspace to begin a new model. A warm start takes you back into the last model you were working in. Click on COLD START at the top right of the screen for a new CATIA session.

5. Depending on your setup, a CATIA News panel may be displayed informing you of recent changes, the most recent level of software, and so forth. If this panel is displayed, press <Enter> to continue to the next stage. You are now logged into a CATIA session and ready to begin.

Work Area

The CATIA screen shown in the next illustration is comprised of numerous elements. At the top of the screen is the pull-down menu bar. To the right are the current palette and current function menu. Located at the bottom of the screen are the fixed menu bar; message area; input information area; prompts; and YES, NO and INTerrupt buttons.

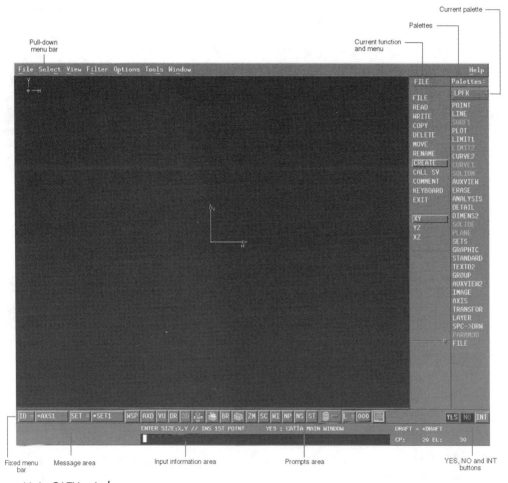

Main CATIA window.

Work Area

After logging on, the session will default to FILE function. In brief, the FILE function will have been automatically selected.

Selections used in Chapters 1 to 3 are discussed in subsequent sections of this chapter. Remaining selections in the main CATIA window are covered as they are introduced in Chapters 4 to 18.

Palettes

Palettes are used to select the desired function and function option. The correct creation and placement of palettes has a substantial impact on the ease and speed of working. The use of an LPFK keypad in conjunction with stored palettes boosts efficiency even further. Although a minimum hardware setup is assumed, references to other hardware items will occasionally be made.

LPFK palette

A general purpose palette (containing File, Erase, Point, Line, Curve2, and Limit1) is extremely useful. A default palette called LPFK will already be stored on your workstation. Nearly all functions required in Chapters 1 to 3 are available on the LPFK palette. Methods to create additional palettes will be presented later as required

Palette placement options described below are accessed by selecting Options from the pull-down menu bar, and then selecting Palette Placement. The four options under Palette Placement are Permanent, Floating, Compact, and None.

• Permanent. From the pull-down menu bar select Options | Palette Placement | Permanent. This option permanently displays both the current palette and the current function menu. The Permanent option is assumed throughout this book.

• Floating. Similar to Permanent, except that the current palette can be placed anywhere in the work area by clicking on the top of the palette and dragging it to the desired position.

• Compact. Only the name of the current palette is displayed along with the current function and respective menu. When the name of the current function is selected, all functions available in the current palette are displayed.

- None. Only the current function and respective menu are displayed.

On the fixed menu bar you will see a toggle button displaying SP or DR. This button is used to change from SPACE mode to DRAW mode and vice versa. SPACE mode is used for working in the 3D environment, and DRAW mode for working in the 2D environment. If the SP option is showing, click on the button to change it to DR. If DR is already displayed, proceed to the next section.

Starting a New Model

In CATIA a model is the name given to the collection of data that you wish to store or file. A new model is started by using the FILE function or the File pull-down menu.

FILE Function

The FILE function should be displayed in the palette area. If the function is not displayed, select FILE at the bottom of the LPFK palette at the right of the screen. From the FILE function menu, select CREATE. (This instruction could also be written as select FILE > CREATE.) In the message area at the bottom of the screen, you will see YES:CONFIRM. Click on the YES button, or press the YES key on the LPFK keypad.

In the center of the work area, note the view axis: a V indicates the vertical axis and an H, the horizontal axis. CATIA is now prepared for the user to begin creating 2D graphic elements. The procedure described in the next section brings the user to the same place.

File Pull-down Menu

Select File from the pull-down menu bar at the top of the screen. In the next illustration, note the words "File Tear-Off" at the top of the pull-down menu. However, when the File pull-down menu appears on the screen, a dashed line runs across the top of the menu. When you click on the dashed line, the menu will be "torn off," that is, it will remain displayed after other selections are made. (In order to produce the illustrations used in this book, certain windows were torn off prior to capture.)

File pull-down menu.

To close a window which has been torn off, select the hyphen (-) symbol at the top left corner of the window and then click on Close. Alternately, you can double-click on the hyphen symbol.

After selecting File, you would choose New to access the New model window. Note that some menu options are followed by an alternative method of execution. In the case of New, the alternative is Ctrl+N.

New model window.

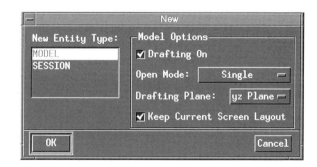

In the New model window, select MODEL in the New Entity Type box. In the Model Options box, a tick should be showing in the small window next to the Drafting On option. If the tick is not displayed, select the window and the tick will appear. (At this stage, window selection is not important because it is related to 3D models. Manipulation of 3D models is covered in Chapter 9 through 18.)

Select the box next to Open Mode and then select Single. Click on the box next to Drafting Plane and then click on YZ plane.

Finally, verify that a tick is showing in the small window next to Keep Current Screen Layout using the method described above. Click on the OK button.

You are now ready to start drawing in a NEW Single model in the YZ view. The significance of the three attributes is discussed below. A view axis is now displayed. The CATIA environment is now prepared for the user to begin producing 2D graphic elements (i.e., draw).

Graphical Representation of 2D Elements

Prior to generating 2D elements (the topic of Chapter 2), you need to choose the graphical representation of the 2D elements you will be creating. The STD (standard) button is used for this purpose. Click on the STD button in the fixed menu bar at the bottom of the screen. A new screen with the heading DRAW ELEMENT GRAPHIC STANDARDS will appear. Selections from this screen set the graphic standard for the elements you create. For Chapters 2 through 8, make the selections below to define the graphic standard.

- POINTS > STAR. To enhance visibility, all created points will resemble asterisks (*).

- LINES > SOLID. All lines will be solid.

- CURVES_SHAPE > SOLID. All curves will be solid.

- THICKNESS > .4000. All lines and curves will be drawn at .4mm thickness.

- COLOR > NONE. All elements will be drawn with no color (white).

- GRID MODE > NO. If YES is displayed, select the word to toggle the mode to NO.

- IMPLICIT POINTS > NO.

- POLAR > NO.

When the above selections are complete, select EXIT from the fixed menu bar.

> **NOTE:** *Whichever method you use to create a new model, use the alternate method the next time you log on.*

2 Generating 2D Graphic Elements

Drawing 2D elements in CATIA is effectively the same as constructing geometry when drawing on paper, except that CATIA provides you with a large number of tools to construct and modify complex geometry more easily and with much more accuracy. In addition to providing direction on how to draw in CATIA, this chapter is focused on how CATIA works. Topics covered include how to input information, how to respond to CATIA commands, and further detail on the CATIA screen layout (palette area, pull-down menus, message areas, and so forth).

This chapter covers the generation of 2D graphic elements—points, lines, and curves—with the use of the following commands from the palette.

- POINT
- LINE
- CURVE2

Alternatives for Creating Graphical Elements

Points, lines, and curves can be created in many different ways. The following tables present available options by function and descriptions of each option.

Chapter 2: Generating 2D Graphic Elements

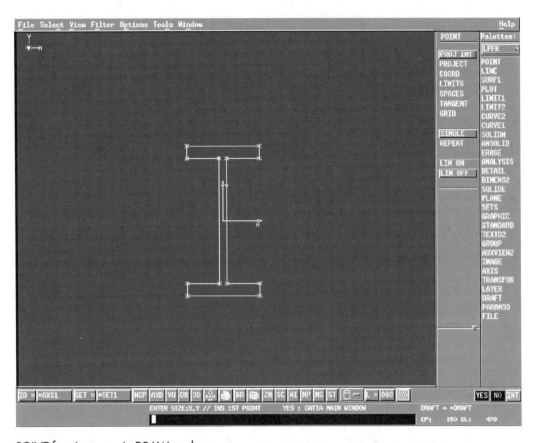

POINT function menu in DRAW mode.

POINT in DRAW Mode

Option	Description
PROJ INT	Select intersecting lines or curves.
PROJECT	Project existing points onto other elements.
COORD	Enter X and Y coordinates, or R theta in polar mode.
LIMITS	Place limit points on lines or curves.
SPACE	Place equidistant points on lines or curves.
TANGENT	Place tangency points on a curve with respect to other elements.
GRID	Create a point grid.

Alternatives for Creating Graphical Elements 13

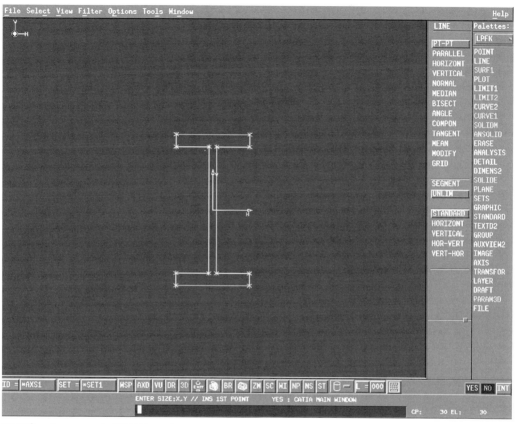

LINE function menu in DRAW mode.

The LINE function is used to create and modify unlimited lines, line segments, and grids of lines.

LINE in DRAW Mode

Option	Description
PT-PT	Select points, ends of existing lines, or curves.
PARALLEL	Use existing lines to create parallel lines.
HORIZONT	Use points to create horizontal lines.
VERTICAL	Use points to create vertical lines.
NORMAL	Use points and lines to create normal lines.
MEDIAN	Use points or a line to create a median line.

Chapter 2: Generating 2D Graphic Elements

Option	Description
BISECT	Bisect existing lines.
ANGLE	Use points or lines to create angled lines.
COMPON	Use the components of a line.
TANGENT	Use points, lines, or curves to create tangential lines.
MEAN	Use at least two points to create a mean line.
GRID	Create a line grid.

Lines can also be modified using the MODIFY option from the LINE function menu.

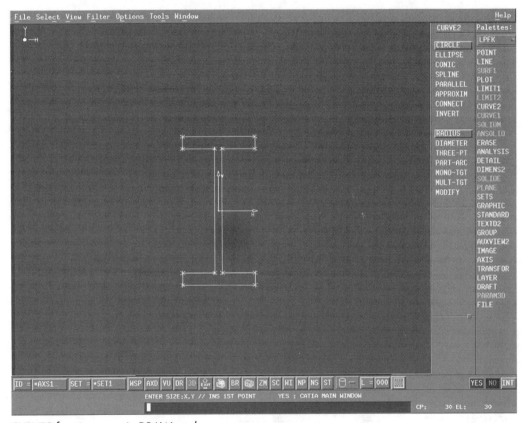

CURVE2 function menu in DRAW mode.

The CURVE2 function is used to create and modify curves (circles, conics, and splines).

CURVE2 in DRAW Mode

Option	Description
RADIUS	Define center and radius to create a circle.
DIAMETER	Define center and diameter to create a circle.
THREE-PT	Define three points or two points and radius to create a circle.
PART-ARC	Define three points or two points and radius to create an arc.
MULTI-TGT	Use one or more elements to create tangent circles where the center point is or is not known.
MONO-TGT	Use one or more elements to create tangent circles where the center is known.
ELLIPSE	Define geometry to create an ellipse.
CONIC	Use three or five points to create conic curves.
SPLINE	Use two or more points to create a spline.
PARALLEL	Use an existing curves (or lines) to create a parallel curve (or lines).
APPROXIM	Use an existing curve to create another curve with a given degree and tolerance value.
CONNECT	Use two curves to create a connecting curve.
INVERT	Invert an existing curve.

Circles can also be modified using the CIRCLE-MODIFY option from the CURVE2 function menu.

Using POINT, LINE, and CURVE2

The following exercises demonstrate the most frequently used options of the POINT, LINE, and CURVE2 functions. Less frequently used options will be introduced in subsequent chapters. Although these initial exercises will not appear to be very useful objects, the shapes serve to familiarize you with CATIA.

Setting Up

Before creating 2D elements, you must verify that your workstation is set up for 2D and that you are working in a new model. The following steps reproduce the procedure discussed in Chapter 1 to clarify the principles involved in setting up for 2D drawing.

1. Select the FILE function at the bottom of the LPFK palette to the right of the screen.

2. In the FILE menu, select CREATE, followed by YZ.

3. In the message area of the screen you will see YES:CONFIRM. Before clicking on the YES button, verify that the SP/DR button on the fixed menu bar displays DR. If DR is not displayed, click on the button to toggle it to DR.

4. Click on the YES button at the bottom right of the screen. You should see a view axis in the center of the work area, where V indicates the vertical axis and H, the horizontal.

Model Manipulation

Before beginning with the first exercise, take a few minutes to experiment with manipulating your model on the screen. Users with access to dials can perform the following manipulations:

- Dial 1 translates the model left or right
- Dial 2 translates the model up or down
- Dial 3 translates the model in or out

Although only the axis is on the screen at the moment, experiment with the dials to view their effects on the axis.

Users without dials can use the keyboard cursor keys to manipulate the model as follows:

- To zoom in use the up cursor key.
- To zoom out use the down cursor key.
- To translate the model to the left use the <Shift> key and the left cursor key.
- To translate the model to the right use the <Shift> key and the right cursor key.
- To translate the model vertically up use the <Shift> key and the up cursor key.

- To translate the model vertically down use the <Shift> key and the down cursor key.

Experiment with the manipulation options to view effects on the axis.

The other option for model manipulation is drag, as discussed in Chapter 1. To drag the model around the screen, position the cursor on the screen, hold down mouse button 3, and move the cursor. The model will follow the cursor around the screen.

If you have dials you will be able to drag using mouse button 3 and dial 3 at the same time. This option is an extremely useful facility for moving around the screen very quickly.

When you zoom in and out using the cursor keys the model is centered in the middle of the screen. But if you position the cursor on the screen and hold down mouse button 3 and then use the cursor keys to zoom in and out, the model will be centered on the cursor position.

Exercise 1: Creating a Rectangle

The objective of this exercise is to create points in the four corners of a rectangle and then to join the points with segment lines. To begin, take the following steps to create the points.

1. To access the point coordinate mode, select POINT > COORD | SINGLE | RECTANGLE. The message area of the screen will display the following: REFERENCE POINT   SEL VU   KEY X,Y   IND POINT.

2. Using the KEY X,Y method of entering the coordinate positions of the four points, the points are entered in the X,Y format in the input information area. Because the bottom left point is on the axis, its coordinate position is 0,0. However, when the X or Y coordinate is zero, you need not key in 0. In this case, key in the comma (,) only, and press the <Enter> key.

3. The next point to enter is at top left, where X=0 and Y=30. Key in *,30* and press <Enter>.

4. The coordinate position of the top right right is X=30 and Y=30. Key in *30,30* and press <Enter>.

5. Coordinates for the last point at the bottom right are X=30 and Y=0. Input *30,* and press <Enter>.

You should now see four points on your screen. The next elements to be created are the joining lines. Take the following steps to create the lines.

1. Select LINE > PT-PT | SEGMENT | STANDARD. The message area will display the following: SEL 1ST PT<LN/CRV>.

> **NOTE:** *Abbreviations such as LN and CRV are explained in detail in Appendix A.*

2. To create the first vertical line, select the point on the axis (bottom left corner), and then the second point created (top left corner). The first line appears.

3. Create the remaining three lines by selecting the pairs of corner points in turn. The drawing should resemble the next illustration.

An alternative method for creating the lines involves the use of LINE > VERTICAL | SEGMENT | ONE LIM and LINE > HORIZONT | SEGMENT | ONE LIM. Take the following steps to experiment with this method.

1. Select LINE > VERTICAL | SEGMENT | ONE LIM.

2. Select the bottom left point.

3. Input the length of the line by selecting the top left point or by keying in *30* to the input information area.

4. The remaining lines can be created in a similar manner using the vertical or horizontal line options (i.e., LINE > VERTICAL... or LINE > HORIZONT...).

Yet another method of creating the lines involves the use of LINE > PT-PT | SEGMENT | HOR-VERT or LINE > PT-PT | SEGMENT | VERT-HOR. Take the following steps to experiment with this method.

Exercise 1: Creating a Rectangle

1. Select LINE > PT-PT | SEGMENT | HOR-VERT.

2. Select the bottom left point and then the top right point. CATIA will draw the horizontal line first and then the vertical line.

3. The remaining two lines can be created using the same function. Select the top right point first and then the bottom left point. Alternatively, the two lines could be created by selecting LINE> PT-PT | SEGMENT | VERT-HOR. At this juncture, you would select the points in the same order as in step 2, or bottom left and then top right.

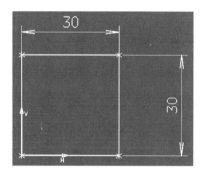

Rectangle with points in each corner.

To save the completed model, take the following steps.

1. Select the FILE function from the palette, and then select the FILE option. The message area displays the following: KEY STRING. If you press <Enter> while leaving the input information area blank, you will be presented with a list of available files. From this list you would select the file in which you wish to store your model.

2. Select the WRITE option from the FILE function menu. Key in a file name (e.g., *CHAPTER2 EXERCISE1*) to the input information area, and then press <Enter>.

3. The message area displays the following: MODEL WRITTEN. The model has been saved.

➥ **NOTE:** *CATIA always writes model file names in upper case, regardless of whether you input upper- or lower-case file names.*

Exercise 2: Adding Lines, Points, and Curves

The objective of this exercise is to add lines, points, and curves to the geometry created in Exercise 1. To create lines from the midpoints of the horizontal and vertical lines, take the following steps.

1. To create the vertical line, select LINE > MEDIAN | SEGMENT | ONE LIM. These selections will permit you to create lines from the midpoint of a selected line. The new line will be normal to the selected one.

2. Select the lower horizontal line anywhere along its length. The midpoint of the line will be highlighted.

3. To define the new line's length, simply select the upper horizontal line. A vertical line from the midpoint of the lower line limited to the upper line will be created.

4. To create the horizontal line, you could use the same option. For the purpose of experimentation, use an alternative. Select LINE > HORIZONT | SEGMENT | ONE LIM. This option allows you to create only horizontal lines.

5. Before using the option in step 4, you need to create a point at the midpoint of the vertical line. To create this point, select POINT > SPACES from the palette. After selecting the left vertical line, the following message displays: KEY NUM //YES:MEDIAN POINT. Click on the YES button at the bottom right of the screen. You have now created a midpoint.

6. Return to the LINE function from the palette by selecting LINE > HORIZONTAL | SEGMENT | ONE LIM.

7. Select the point just created on the vertical line and then select the opposite vertical line. The horizontal line has been created.

The next part of this exercise is to create the smaller of two circles.

1. To create the point on the intersection of the two newly created lines, select POINT > PROJ INT | SINGLE | LIM ON. Select each of the newly created lines in turn; the new point is created.

2. Select CURVE2 > CIRCLE | RADIUS. To define the circle center, select the point created in step 1.

Exercise 2: Adding Lines, Points, and Curves   21

3. In the message area you are asked to key in the radius, and select another point or line to define the radius of the circle. In the input information area, input 5 and press <Enter>. The first circle is created.

To create the second circle, take the following steps.

1. Select CURVE2 > CIRCLE | MULTI-TGT | UNSPEC. The message area displays the following: SEL 1ST PTD/LND/CIRD/CRVD.

2. Select any one of the outer lines. The message area displays the following: SEL 2ND PTD/LND/CIRD/CRVD.

3. Select one of the remaining lines. The message area now displays SEL 3RD PTD/LND/CIRD/CRVD KEY RAD//IND REGION.

4. Choose one of the two remaining lines, and a circle will be created. The message area displays YES:NEXT. This message refers to the fact that for any three elements selected to create a tangent circle there will be more than one solution. To view the other solutions, simply click on the YES button at the bottom right of the screen and CATIA will scroll through all available solutions. The second drawing is complete.

5. Save this drawing in the same way as described at the end of Exercise 1. Give it a new name such as *chapter2 exercise2*.

Lines, points, and curves added to geometry created in Exercise 1.

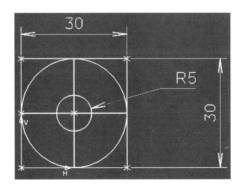

Exercise 3: Creating an Angle Section

The objective of this exercise is to create a drawing of an angle section. To begin, take the following steps to create a new blank model.

1. Select the FILE function from the bottom of the LPFK palette. Select CREATE | YZ. The message area displays YES:CONFIRM.

2. Prior to clicking on the YES button, verify that the SP/DR button on the fixed menu bar displays DR. If the button displays SP, toggle the button to display DR. Click on the YES button at the bottom right of the screen.

In the center of the work area, you should see a view axis. The V indicates the vertical axis, and the H, the horizontal axis. You are now ready to begin.

1. Create the corner points by selecting POINT > COORD | SINGLE | RECTANG.

2. Key in the following corner coordinates in the input information area: bottom left corner *0,0*; top left corner *0,30*; top right corner *5,30*; middle left corner *5,5*; middle right corner *30,5*; and bottom right corner *30,0*.

3. To join the corner points, create the sides of the angle section using LINE >PT-PT | SEGMENT | ONE LIM.

4. Save the drawing.

Equal angle section.

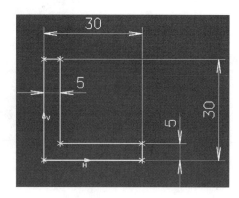

Exercise 4: Creating an I Beam

The objective of this exercise is to create a drawing of an I beam. The method of creating the I beam is the same employed in Exercise 3.

1. Create a new blank model. Select FILE > CREATE | YZ, followed by YES to confirm creation.

2. Create the points. Select POINT > COORD | SINGLE | RECTANG, followed by the coordinates for all corner points.

3. Create the lines. Select LINE > PT-PT | SEGMENT | ONE LIM, followed by selecting pairs of points to create the lines.

4. Save the drawing.

◦ NOTE: *Similar to Exercise 1, the I beam could have been created in many other ways. Consider returning to Exercise 1 and experimenting with alternate methods of creating the I beam.*

I beam.

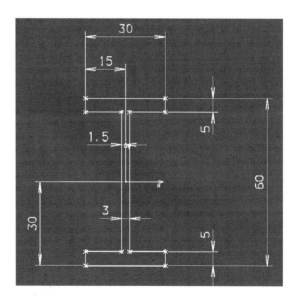

Summary

This chapter covered a variety of basic tasks, including the creation of a new model, creating points with the use of coordinate dimensions, and creating points on the intersection of elements. Additional tasks were creating lines using point to point limits, and the LINE function's vertical, horizontal, vertical-horizontal, horizontal-vertical, and component options. Next, circles were created with the use of points to define their centers, and the use of lines to define tangent circles. Finally, the chapter described how to save and read models from a file.

The drawings created in Exercises 3 and 4 will be used in Chapter 3 for experimentation with modifying and erasing elements.

3 Modifying 2D Graphic Elements

This chapter is focused on modifying points, lines, and curves with the following functions.

- ERASE
- LIMIT1
- GRAPHIC
- MARKUP

The above functions are used to erase elements that have been incorrectly created or are no longer required, make elements longer or shorter, and change the graphical representation of elements, such as in altering line type, color, or thickness. The POINT, LINE, and CURVE functions discussed in Chapter 1 will also receive attention here in order to demonstrate CATIA's versatility.

Alternatives for Modifying Graphical Elements

Graphic elements can be modified in many different ways. The following tables present available options according to ERASE, LIMIT1, MARKUP, and GRAPHIC functions.

Chapter 3: Modifying 2D Graphic Elements

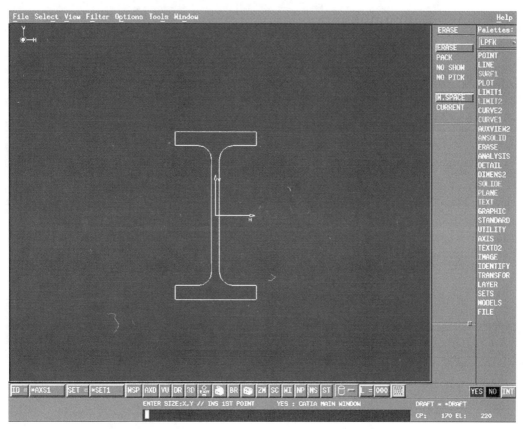

ERASE function menu in DRAW mode.

ERASE in DRAW Mode

Option	Description
ERASE	Delete elements from the model.
PACK	Pack the model after element deletion.
NO SNOW/SHOW	Hide elements from view, and retrieve elements that have been hidden.
NO PICK/PICK	Make elements nonselectable and retrieve elements that have been made nonselectable.

Alternatives for Modifying Graphical Elements

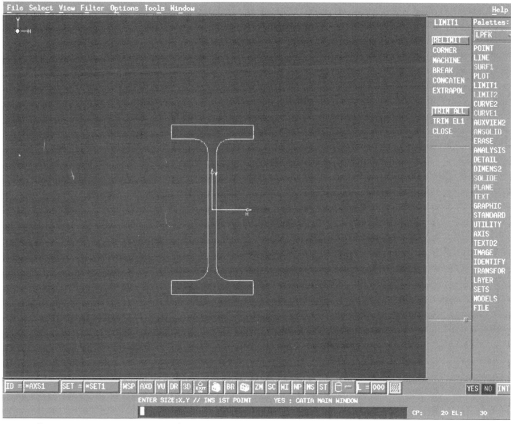

LIMIT1 function menu in DRAW mode.

The LIMIT1 function is used to modify lines, circles, conics, and curves. LIMIT1 can also be used to create connecting elements (e.g., chamfers and fillets).

LIMIT1 in DRAW Mode

Option	Description
RELIMIT	Use existing lines, curves, and points to relimit existing lines or curves.
CORNER	Add connecting curves to existing lines and curves.
MACHINE	Add chamfers to existing lines and curves.
BREAK	Break lines and curves.
CONCATEN	Join two lines or curves to create a single element.
EXTRAPOL	Extend the length of a line or curve.

Chapter 3: Modifying 2D Graphic Elements

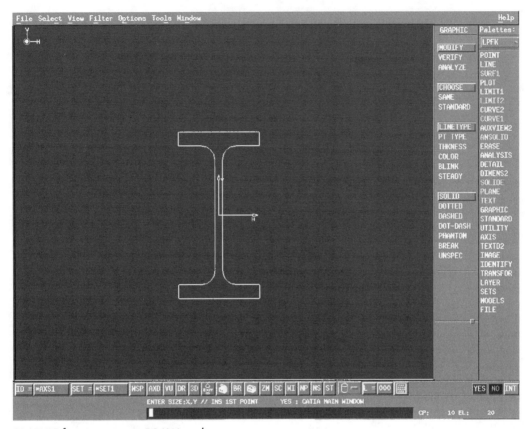

GRAPHIC function menu in DRAW mode.

The GRAPHIC function is used to modify the graphical representation of lines, curves, and points.

GRAPHIC in DRAW Mode

Option	Description
LINETYPE	Change line type (e.g., solid, dotted).
PT TYPE	Change point type (e.g., star, cross).
THKNESS	Change line thickness.
COLOR	Change color.
BLINK/STEADY	Highlight elements.

The graphical attributes used for any element can also be checked using the ANALYZE option from the GRAPHIC function menu.

Alternatives for Modifying Graphical Elements

The MARKUP function is used create center lines, screw threads, and various arrow types for use in annotating drawings.

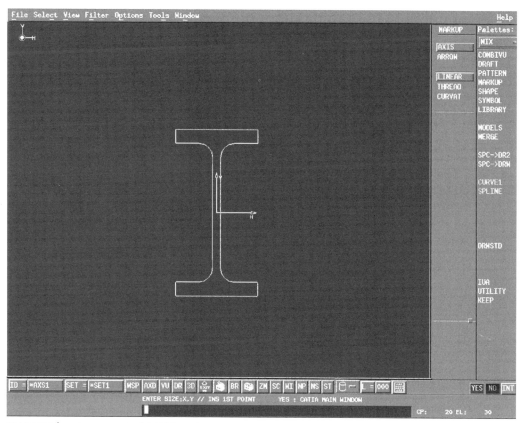

MARKUP function menu in DRAW mode.

MARKUP in DRAW Mode

Option	Description
AXIS I LINEAR	Use existing lines and curves to create center lines.
AXIS I THREAD	Use an existing circle to create a concentric broken circle to symbolize a screw thread. The broken circle can be inside or outside the selected circle to enable symbolization of male or female threads.
AXIS I CURVAT	Use circles on a pitch circle to create the center line running through them relative to their respective centers.
ARROW	Use existing elements or indicate positions to create arrows for use in annotating drawings.

The following exercises focus on the most frequently used options for the four functions. Prior to working with the exercises, verify that your workstation is set up for 2D and that you are working in a new model. (If necessary, see Chapter 1 for a review.) The first exercises are duplicated from Chapter 2 in order to introduce the LINE function's unlimited line options. Alternate methods to create target drawings are discussed. Although the objects created in the exercises will not appear to be extremely useful, the shapes serve as an introduction to using CATIA, how the program communicates with you and vice versa.

Exercise 1: Creating a Simple Box Using Unlimited Lines

The objective of this exercise is to create a simple box using unlimited lines. LIMIT1 is then used to "relimit" the lines.

1. To create a vertical line through the axis, select LINE > VERTICAL | UNLIM. Because the first vertical line must pass through the axis, a distance entry is not necessary. Press <Enter>.

2. The second vertical line to be created should be at a distance of 30 from the vertical axis. Type *30* in the input information area, and press <Enter>.

3. To create the two horizontal lines, select LINE > HORIZONT | UNLIM. The first line can be created by simply pressing <Enter>. For the second line, enter *30* in the input information area and press <Enter>.

Before continuing to the limit procedure, the creation of the four lines will be examined. The first vertical and horizontal lines have no distance from the vertical and horizontal axis, respectively. Consequently, by pressing the <Enter> key you are instructing CATIA to simply create lines that pass through the axis. The next two lines are both at a distance of 30 from the vertical and horizontal axis, respectively. Therefore, by entering *30* you are instructing CATIA to create lines which are parallel to the vertical and horizontal axes at a distance of 30 from the axes.

Exercise 2: Exploring ERASE Function Options

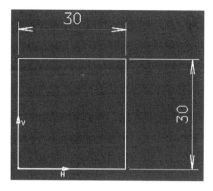

Simple box.

To limit the four lines, take the following steps.

1. Select LIMIT1 > RELIMIT | TRIM ALL.

2. Select the first vertical line somewhere between the two horizontal lines, and then select the upper horizontal line somewhere between the two vertical lines. You have now limited the first corner.

3. To limit the remaining corners, repeat step 2.

4. Save the drawing. (See Exercise 1 in Chapter 2 if you are not sure about how to proceed.)

The second vertical and horizontal lines could also have been created by selecting LINE > PARALLEL | UNLIM. With this method the second vertical line would be created by selecting the first vertical line, indicating (via mouse button 2) the side on which you require the line to be, keying in *30* to the input information area, and pressing <Enter>. The second horizontal line could then be created using the same option followed by selection of the horizontal line.

Exercise 2: Exploring ERASE Function Options

Read the results of Chapter 2, Exercise 1, and take the following steps.

1. Select ERASE > ERASE | W.SPACE. Select an element. If you inadvertently select an element that you do not wish to erase, click on the NO button at the bottom right of the screen.

✗ **WARNING:** *You can retrieve only the last element selected, and only if you have not changed function or option.*

2. To hide an element from view, select ERASE > NO SHOW | W.SPACE. If you inadvertently select an element that you do not wish to hide, you can retrieve the element by clicking on the NO button at the bottom right of the screen.

3. Note that the message area displays YES:SWAP. If you click on the YES button at the bottom right of the screen, you will swap into the hide area where you can view elements previously selected to be hidden. Next, the NO SHOW option in the menu will change to SHOW. Therefore, to return hidden elements to visibility, select such elements, and click on the YES button to return to the show area.

4. Select ERASE > NO PICK | W.SPACE. To make an element nonselectable, simply select it with mouse button 1. This element will display, but can no longer be selected. When you inadvertently select an element that you do not wish to make nonselectable, retrieve it by clicking on the NO button at the bottom right of the screen.

5. Note that the message area displays YES:SWAP. Click on the YES button to swap into the nonselectable area to view elements marked as nonselectable. Note that the NO PICK option in the menu has changed to PICK. To return nonselectable elements to selectable status, pick them. The message area continues to display YES:SWAP. Click on the YES button to return to the show area.

6. Do not save the altered Exercise 1 file.

Exercise 3: Creating an Angle Section with Radius Corner

The objective of this exercise is to create a drawing of an angle section, but with a radius in its corner instead of a sharp corner. Before taking the following steps, create a new model.

Exercise 3: Creating an Angle Section with Radius Corner

1. To create the first line (the vertical line passing through the axis), select LINE > VERTICAL | UNLIM and press <Enter>.

2. To create the horizontal line passing through the axis, select LINE > HORIZONT | UNLIM and press <Enter>.

3. In Chapter 2, Exercise 1, you used the VERTICAL and HORIZONT options to create the four lines. For this exercise, use the PARALLEL option by selecting LINE > PARALLEL | UNLIM. To create the other two vertical lines first, pick the unlimited vertical line. To indicate where the new vertical line will be placed, use mouse button 2 to click in any location to the right of the first vertical line.

4. At this juncture, you need to inform CATIA of how far away you want the new line. Key in *5* to the input information area, and press <Enter>. The second vertical line is in place.

5. The third vertical line can be created in the same manner as the second one. Use a dimension of *30* instead of *5*.

6. The horizontal lines can be created in the same manner. Select LINE > PARALLEL | UNLIM. The outer five corners will be limited to achieve square corners as in Exercise 1. Pick LIMIT1 > RELIMIT | TRIM ALL, and select the lines on both sides of each corner.

7. To relimit the inner corner with a 5 radius, select LIMIT1 > CORNER | TRIM ALL. Key in *5* to the input information area, and press <Enter>.

8. Select in turn the two lines that form the inner corner in order to create a radius in the corner instead of a sharp corner. The message area displays YES:CONVEX. If you click on the YES button, the corner radius will change from concave to convex, and YES:LINE will appear in the message area.

9. If you click on the YES button, the corner radius will change from convex to a straight line between the endpoints of the lines, and YES:CONCAVE will appear in the message area. Clicking on YES will return the concave corner radius.

10. Save the model.

Equal angle section (with radius corner).

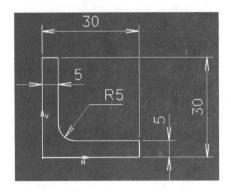

Exercise 4: Creating an I Beam with Radius Corners

The objective of this exercise is to create a drawing of an I beam as in Chapter 2, Exercise 4. However, this time the radius of the inner corners will be 5. The method for creation of the I beam is similar to creating the angle section in the previous exercise. Create a new blank model before taking the following steps.

1. Create all lines using LINE > PARALLEL | UNLIM.

2. Relimit the outer four corners using LIMIT1 > RELIM | TRIM ALL. The model should resemble the next illustration.

I beam with unlimited lines.

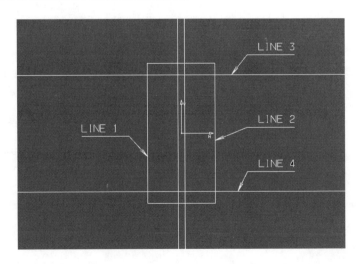

Exercise 4: Creating an I Beam with Radius Corners

At this point, you need to be able to use each of the lines to create the corner radii and the inner sharp corners twice. You could relimit the existing lines you have and then create more lines. The next steps use an alternate method.

1. Select LIMIT1 > BREAK.

2. Pick line 1 and then move the cursor to the approximate midpoint of the line. Use mouse button 2 to indicate a break position. The line has now become two lines and you can use the RELIMIT option to create the two inner left sharp corners.

3. Select line 2 and break it in the same manner as in steps 1 and 2. After breaking this line, continue by breaking lines 3 and 4.

4. Once the four lines have been broken, create the two inner sharp edges at the right using the RELIMIT option.

5. Create the corner radii using the CORNER option with the radius set at 5. The drawing should resemble the first illustration in this section.

6. Save the drawing.

I beam with radius corners.

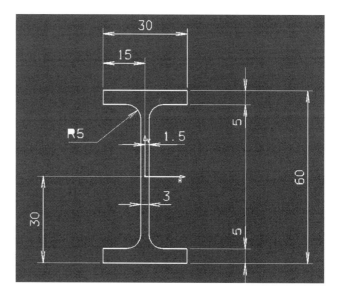

If you had wished to place chamfers on the inner corners, you could use the MACHINE | CHAMFER options of the LIMIT1 function. To experiment with the chamfers, erase the radius corners in the previous model using ERASE > ERASE | W.SPACE, and then select LIMIT1 > MACHINE | CHAMFER | ANGLE | TRIM ALL. To create the chamfer, select two adjacent lines to form the corner; KEY LNG (,ANG) // YES:STD appears in the message area.

The abbreviation LNG is asking you to key in the length for the chamfer leg. ANG requests the angle for the chamfer; it is shown in parentheses to indicate that keying in an angle is not necessary if the angle you require is 45. The last part of the message, YES:STD, means that if you want to use the same leg length and angle as the previously created chamfer, click on the YES button. The previously created chamfer dimensions will be shown in the message area.

Exercise 5: Creating an I Beam with Dot-dashed Center Lines and Hidden Axis

1. Read the model saved at the end of Exercise 4. To create the center lines, you could use LINE > VERTICAL | UNLIM for the vertical line and LINE > HORIZONT | UNLIM for the horizontal line.

2. Limit the lines as indicated in the previous illustration by selecting LIMIT1 > RELIMIT | TRIM ALL. Earlier in the chapter, one line was limited to another. In this instance, the line will be limited to an arbitrary position. Select the vertical line at a point near the axis with mouse button 1. Move the cursor to just above the top horizontal line and indicate using mouse button 2. The line will be shortened to the indicated position.

3. To limit the bottom of the line, select the vertical line near the axis. Move the cursor below the bottom horizontal line and indicate a position to which the line will be limited.

4. Now that the vertical line has been limited, you need to limit the horizontal line. Select the horizontal line near the axis and indicate to the right side of the I beam. Now select the horizontal line again near the axis and indicate to the left side of the I beam.

Exercise 5: Creating an I Beam with Dot-dashed Center Line   37

5. To hide the axis, select ERASE > NO SHOW | W.SPACE and select the axis. The message area displays YES:SWAP.

🔹 **NOTE:** *When you wish to return an element to visibility, click on the YES button to swap into the hide (NO SHOW) area of the model. At this point you would select the element(s) to be made visible.*

6. To change the two center lines from solid to dot-dashed lines, select GRAPHIC > MODIFY | CHOOSE | LINETYPE | DOT-DASHED. (The line type can also be selected from the line type table.) Select the vertical center line and then the horizontal center line. The lines will change from solid to dot-dashed.

7. Save the model.

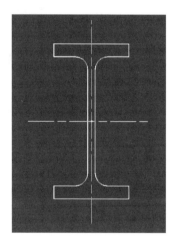

I beam with dot-dashed center lines and hidden axis.

🔹 **NOTE:** *The six standard line types are SOLID, DOTTED, DASHED, DOT-DASHED, PHANTOM, and BREAK. Selecting the UNSPEC option will display 32 line types to choose from. The list is dependent on your configuration, but the first six line types are always the same. The next illustration is a sample of line type alternatives.*

Line type table.

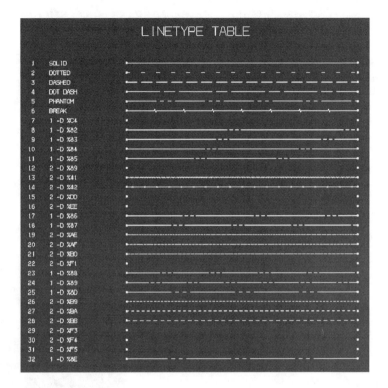

Use the model you have just saved to experiment with line types.

The next option available in GRAPHIC after LINETYPE is PT TYPE. To be able to see the effects of different point types you first need to create some points in the model. Consider using some of the options from the POINT function that you have not yet used.

1. The first option you could use is POINT > LIMITS. With this option, select the top horizontal line. A point appears at both ends of the line.

2. Use POINT > SPACES. With this option select the top horizontal line again; KEY NUM // YES:MEDIAN POINT appears in the message area. KEY NUM is instructing you to key in the number of points you require along the line, YES:MEDIAN POINT means click the YES button if you require only a single point in the middle of the line.

Exercise 5: Creating an I Beam with Dot-dashed Center Line

3. Key in *2* to the input information area and the following message appears in the message area: YES:USE END POINTS. This means click the YES button if you require the endpoints of the line to define the limits of the spacing. In this case go ahead and click the YES button. You will now have two points equispaced along the line.

4. Now that you have some points in the model you can proceed to examine point type options in GRAPHIC. Select GRAPHIC > MODIFY | CHOOSE | PT TYPE. You have a choice of the following point types: Dot, Plus, Cross, and Star. Select the different types in turn and change the points in your model to view the effect.

5. Select GRAPHIC > MODIFY | CHOOSE | THKNESS. Having selected the above function and options you will note that you have a choice of six different thicknesses for use on elements: 1 - 0.1mm or 0.0039" Thickness; 2 - 0.2mm or 0.0079"; 3 - 0.4mm or 0.0157"; 4 - 0.6mm or 0.0236"; 5 - 0.8mm or 0.0315"; and 6 - 1.0mm or 0.0394". (All the exercises in this book are set to mm.) Select the different types in turn and change the lines in the model to view the effect. If the thickness option is switched off in the Display and Manipulation Function Menu you will see only a change to visualization when you change a line to thickness 1. The element chosen will be shown dimmed. These thickness settings are used primarily for plotting drawings.

> **NOTE:** *The Display and Manipulation Function menu is activated by using one of the following methods: pressing the <F4> key; pressing button 4 on a four-button puck; or holding down button 3 and pressing button 1 on a three-button mouse.*

6. The next option available in GRAPHIC after LINETYPE, PT TYPE, and THKNESS is COLOR. Select GRAPHIC > MODIFY | CHOOSE | COLOR. The color modification option is also available from the Tools pull-down menu. Having selected the above function and options, note that when a new window appears on the screen, the window allows you to select any of the available colors. The first five colors in the window are the primary colors: 1 - White; 2 - Red; 3 - Green; 4 - Blue; 5 - Yellow. To use a particular color simply select it from the window, and then select the elements that you want changed to that color. Select a few colors in turn and change the lines in the model to view the effect.

✗ **WARNING:** *Be very careful if you select color number 6, or indeed any color near 6, because you will not be able to see it on your screen. The screen color itself is normally set to 6.*

7. The next two options available in GRAPHIC are BLINK and STEADY. Select GRAPHIC > MODIFY | CHOOSE | BLINK. The effect of this option depends on your CATIA hardware setup. If you are running 5080 CATIA, any element selected under this option will blink, but if you are running Motif CATIA or graPHIGS CATIA, any element will simply be highlighted.

8. Select GRAPHIC > MODIFY | CHOOSE | STEADY. If you selected elements with BLINK, selecting them when using this option will return the elements to their normal state.

9. If at any time you are uncertain of the graphical attributes of a particular element you can analyze the element using the ANALYZE option. Select GRAPHIC > ANALYZE. If you now select any line element you will be informed of its graphical attributes (Color, Line type, and Line thickness). These attributes will be displayed in the alphanumeric screen. Additional information on how to use this screen appears in Chapter 4.

Exercise 6: Creating a Universal Beam

This exercise is similar to the I beam creation exercise earlier in this chapter, but is based on a realistic beam size rather than convenient dimensions. While the following steps to create the beam are recommended in this instance, recall that there are many different ways to create a universal beam shape.

1. Create all the horizontal lines using LINE > HORIZONT | UNLIM. The top line would be created by keying *103.4* to the input information area and pressing <Enter>. This number could also have been keyed in as *206.8/2*; CATIA recognizes this as a function and will perform the calculation for you. All other horizontal lines can now be created in the same manner. Recall that all lines below the axis must be entered as negative numbers.

Exercise 6: Creating a Universal Beam

2. Create all the vertical lines using LINE > VERTICAL | UNLIM. The line at the right would be created by keying *66.9* or *133.8/2* into the input information area and pressing <Enter>. All other vertical lines can be created in the same manner. Recall that all lines to the left of the axis must be entered as negative numbers.

3. Relimit the outer corners using LIMIT1 > RELIMIT | TRIM ALL.

4. To relimit the inner corners, break all the lines forming these corners using LIMIT1 > BREAK. Select the line to be broken and then indicate (with mouse button 2) where the break should be positioned.

5. Relimit the remaining corners using LIMIT1 > RELIMIT | TRIM ALL and LIMIT1 > CORNER | TRIM ALL. Key in a radius of *7.6*.

6. Save the model. This model will be used in subsequent chapters.

203 x 133 universal beam.

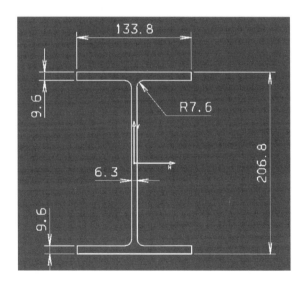

Exercise 7: Creating a Side Plate Using Additional Options

The model in this exercise lacks dimensions in order to introduce additional options of the POINT, LINE, and CURVE2 functions. Consequently, the exercise is substantially different from previous tutorial walkthroughs.

1. To create the top horizontal line, select LINE > HORIZONT | UNLIM. Because this line passes through axis, you do not need to enter a number to the input information area before pressing <Enter>.

2. The next line to be created is the bottom horizontal line. Select LINE > HORIZONT | UNLIM. Key in *-172* to the input information area and press <Enter>.

   Theoretically, you could create the two vertical lines next. However, you need to calculate the position of the lines first because you do not have the dimensions to create the lines. Thus, we will proceed by creating the top two corners because the locations of the corner centers are known.

3. Create the vertical lines on which the corner centers lie. Select LINE > VERTICAL | UNLIM. Key in *99* and press <Enter> to create the right side line, and then key in *-99* and <Enter> to create the left side line.

4. At this point, a CURVE2 option is used to enable you to create a circle when you have the line on which the center lies, the line tangent to the circle, and the radius. Select CURVE2 > CIRCLE | MONO-TGT, and pick the right side line 99 from the vertical axis (line on which the circle center lies).

5. Select the top horizontal line; the circle is tangent to this line. Key in *25*, the radius of the circle, and press <Enter>. Indicate the position of the circle below the top horizontal line. A circle should be located at the top right of the drawing. You can now create a similar circle on the left side.

6. To create the two vertical lines, select LINE > VERTICAL | UNLIM. You do not have to enter dimensions because all the information you need to create the lines is already on the screen. To create the right side line,

Exercise 7: Creating a Side Plate Using Additional Options

select the circle on the right side. A vertical line tangent to the circle appears. Follow the same procedure to create the left side vertical line.

NOTE: *If you select the circle in the wrong position, you may inadvertently create a line on the wrong side of the circle. Click on the NO button to delete the line, and reselect the line.*

The next elements to be created are the four circles. Typically, you would create circles by selecting a center point and keying in a radius. In this exercise, however, the circles will be created by defining the centers with coordinate dimensions.

1. The first circle to be created will be at the top right side. Select CURVE2 > CIRCLE | DIAMETER.

2. To define the center of the circle, key in the X,Y coordinates of the center, *74,-50*, and press <Enter>. When asked to input the diameter, key in *21*, and press <Enter>.

3. The other three circles can be created in the same manner. When creating the remaining circles, you will not have to key in the diameter. CATIA gives you the option of clicking the YES button as an indicator that you wish to use the standard radius (i.e., the value of the last diameter used). Use the following coordinates: top left circle, -74,-50; bottom left circle, -30,-147; and bottom right, 30,-147.

The circles were created without first creating points on respective centers. Because the center points of the circles will be required in an exercise in Chapter 5, you now need to create these points.

Select POINTS > LIMITS and select each of the circles in turn. Two points will be created on each circle—one on the circle center and the other on the start and finish points of the circle. The points on the start and finish points may be erased because they will not be required in future exercises.

The next elements to be created are the bottom two corner circles.

1. Select CURVE2 > CIRCLE | RADIUS.

2. Because the centers of two existing circles are in the same positions as the two circles to be created, the existing circles will be used to define

the centers of the new circles. To define the center of the right side circle, select the existing bottom right circle. Key in *25* and press <Enter>.

3. The left side circle can now be created by selecting the bottom left circle. Click the YES key to accept the last used radius.

The final elements to be created are the two angled lines.

1. Select LINE > TANGENT | UNLIM. Pick the bottom right 25 radius circle. When asked for an angle to define the tangency line, key in *45* and press <Enter>. The line is now angled at 45 and tangent to the circle.

2. Create the left angled line using the same method as indicated in step 1. Key in *-45* as the angle.

The final part to the current exercise is to relimit all corners. Start at the top right corner and work clockwise around the drawing.

1. Select LIMIT1 > RELIMIT | TRIM ALL.

2. Pick the top horizontal line to the left of the corner circle and then select the corner circle at the top right. To complete this corner, select the circle in the same place again and then select the right vertical line.

3. Proceed to the next corner. Select LIMIT1 > CORNER | TRIM ALL, key in *25* for the radius, and press <Enter> before selecting any lines.

4. Select the vertical line in the region of the line you wish to retain. Then pick the angled line, again in the region that you wish to retain. You have now relimited the first two corners and the other corners are simply reproductions of these two corners. For the corners where an additional radius is not required, use LIMIT1 > RELIMIT | TRIM ALL. For the corner that requires an additional radius, use LIMIT1 > CORNER | TRIM ALL.

➥ **NOTE:** *When using LIMIT1 > RELIMIT, CORNER, or MACHINE, trimming the wrong side of elements is common. You can usually remedy the situation without having to redo the relimit. After selecting the elements to be trimmed on the wrong side, indicate (with mouse button 2) in another quadrant of the corner. The elements will flip around.*

Exercise 7: Creating a Side Plate Using Additional Options 45

Options available when using relimiting line function.

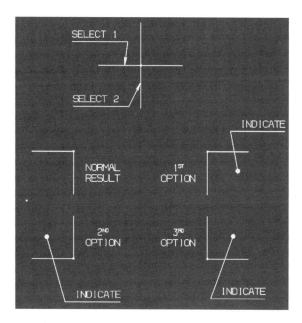

The only lines remaining to create are the four circle center lines. You could create these lines by using LINE function options, relimiting the lines, and changing the thickness of the lines as well as line type to DOT-DASHED. But there is a much simpler way of accomplishing these tasks.

The MARKUP function allows you to create center lines for curves simply by selecting the curve. These lines are created in DOT-DASHED line type and with a thickness of 0.1mm or 0.0039". The only option you have to consider is by how much the center lines will overrun the curve.

Creating Palettes

Most of the CATIA functions required in this book are on the LPFK palette. The few exceptions include MARKUP. To create MARKUP (or any additional palette), access the Palette Creation screen by selecting FILE > KEYBOARD, or click on the Keyboard button at the right end of the fixed menu bar at the bottom of the screen.

Chapter 3: Modifying 2D Graphic Elements

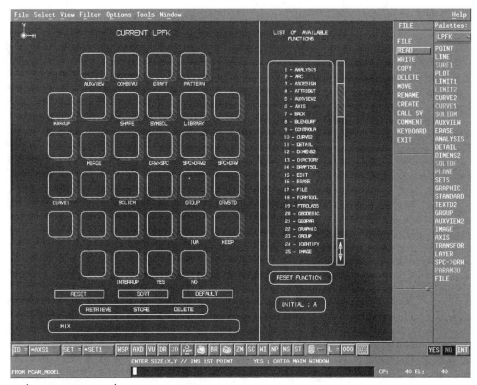

Palette Creation window.

The keyboard area is referred to as the CURRENT LPFK because it is laid out like the LPFK keyboard described in Chapter 1 in the section focused on additional hardware items.

First, you need to reset to produce a blank palette. This is done by clicking on the RESET button. Note that each of the keys is now blank and a list of available functions appears at the right.

To allocate a function to a key, select it from the list, and then click on the key on which you wish to place it. The previous illustration shows a palette with all functions not included on the LPFK palette.

Using the method described above allocate the functions to the keys. When complete, click on the STORE button and then type in a name for the new palette. The suggested name is *MIX*; this palette will henceforth be referred to as MIX.

Exercise 7: Creating a Side Plate Using Additional Options 47

> **NOTE:** *If the desired function cannot be seen on the list, scroll up or down using the double arrow at the lower right of the list panel.*

To exit from the screen, select another function, or click on the EXIT button on the fixed menu if you used the second method to access the Palette Creation screen.

To switch between palettes, click on the current palette name and select the new one from the list displayed. Now that the required palette has been created, you can continue with the exercise.

1. Select MARKUP > AXIS. When CATIA asks for the line overrun, key in 5 and press <Enter>. Select the circle. The result is neat center lines overruning the circle by 5mm and created in dot-dashed line type with a thickness of 0.1mm or 0.0039".

2. For the remaining three circles, select the circle. CATIA will again use the last entered overrun dimension for subsequent circles. (Other MARKUP options are covered in subsequent chapters, but mainly in Chapter 7, "Annotating Drawings.")

3. Save the model. This model will be used in later chapters.

Side plate.

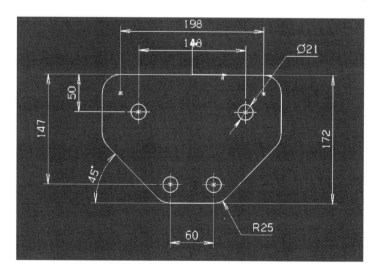

Summary

This chapter focused on several topics, including erasing elements from a model, hiding elements in a model, how to make elements nonselectable, modifying element lengths, adding radius and chamfered corners to model geometry, changing the graphical representation of elements in a model, adding center lines to geometry without having to create lines, and changing the graphical representation of center lines.

Chapter 4 covers analysis geometry created in previous exercises. As you proceed through the chapters, only new functions and options are explained in detail. If you need a refresher in the use of functions discussed in earlier chapters, please review pertinent sections.

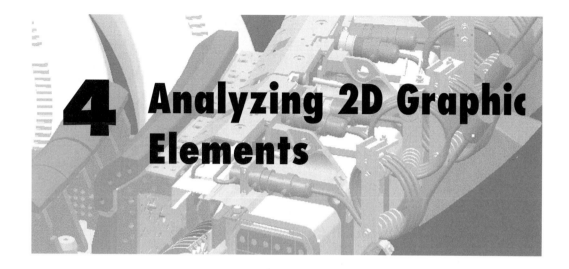

4 Analyzing 2D Graphic Elements

This chapter is focused on analyzing the numerical values of DRAW type elements, as well as the relative values and logical links between DRAW elements (e.g., associativity). After creating 2D elements, you may require information about the elements, such as line length, and the position of line or curve intersections. Next, the area or inertial characteristics of a shape may be important to the design objective. Appearing below are descriptions of options for the ANALYSIS and SHAPE functions in DRAW mode.

ANALYSIS in DRAW Mode

Option	Description
NUMERIC	Analyze the numerical values of DRAW type elements.
RELATIVE	Analyze the relative values of DRAW type elements.
LOGICAL	Analyze the logical links between DRAW elements.

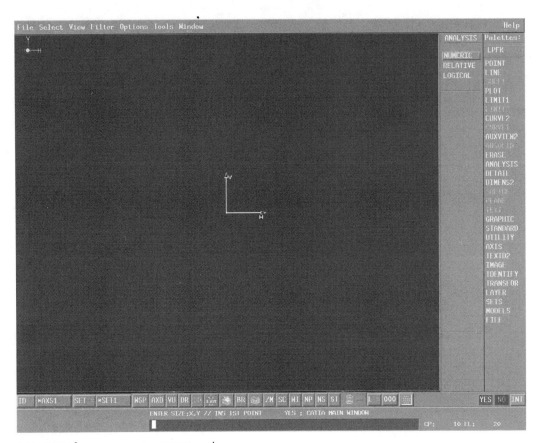

ANALYSIS function menu in DRAW mode.

SHAPE in DRAW Mode

Option	Description
CREATE > POLYGON	Create polygonal domains called shapes which can be closed, square, rectangular, open, or unspecified.
OFFSET	Create a shape parallel to another shape.
MODIFY	Modify the geometry of a shape by moving, adding, or erasing vertices.
OPERATE	Perform Boolean operations on shapes by the intersection, union, subtraction, or splitting of shapes. (Boolean operations will be explained in more detail in later chapters.)
CLOSE	Close an open shape.

Exercise 1: Introduction to Alphanumeric Window

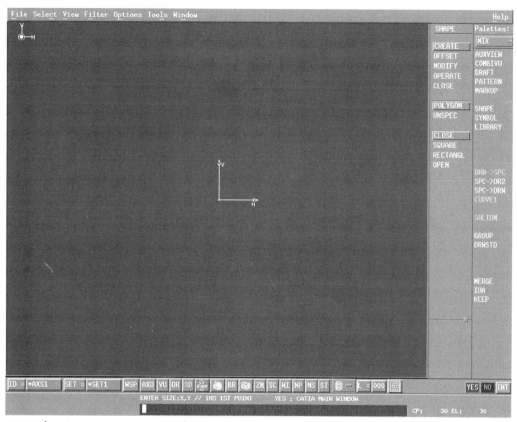

SHAPE function menu in DRAW mode.

In the following exercises, the I beam shape created and saved in Chapter 3 will be used. The ANALYSIS function is employed to obtain numerical and relative information about the geometry produced, and the SHAPE function is used for analysis of the cross section.

Exercise 1: Introduction to Alphanumeric Window

1. Access the I beam shape saved in Chapter 3 using FILE > READ.

2. Select the ANALYSIS function from the LPFK palette. The alphanumeric window appears, which is where analysis results will be displayed.

3. Select ANALYSIS > NUMERIC. The prompt in the message area will tell you to SEL:ELEM (select an element).

4. Select the top horizontal line of the I beam geometry. The following illustration shows the contents of the alphanumeric window.

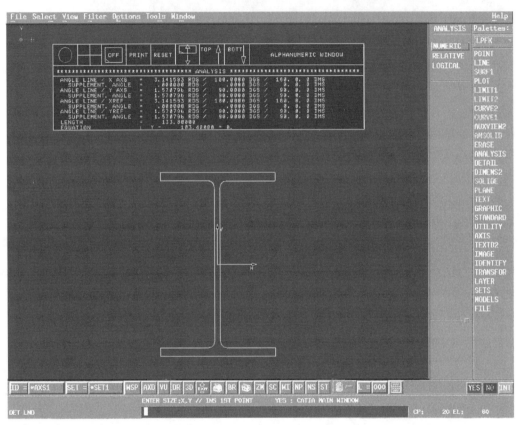

Alphanumeric window in ANALYSIS function.

If the window does not resemble the illustration, click on the circle in the window's top left corner and experiment with the various controls shown at the top (i.e., PRINT, RESET, and so forth).

The alphanumeric window is now the active window. You can drag it using the method described in Chapter 1. This window can be switched on and off manually: press the <Alt> key and the <+> key to switch it on, and <Alt> plus the <-> key to switch off.

Exercise 2: Relative Analysis 53

The window will automatically switch off when another CATIA function is selected. If you wish to manually make the work area active again, select the red circle on the axis icon at the top left corner of the screen.

Exercise 2: Relative Analysis

Let's try some relative analysis. You could choose any two geometric elements, but for the sake of this exercise, you will identify the relationship between the top and bottom of the I beam.

1. Select ANALYSIS > RELATIVE. Pick the top and then the bottom horizontal lines of the I beam shape.

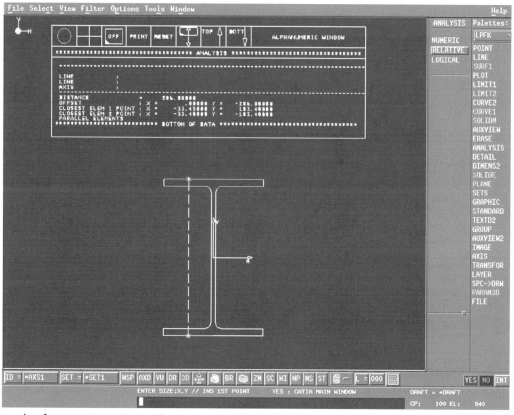

Results of ANALYSIS > RELATIVE.

2. Using the two analysis options above, experiment by making various selections from the I beam shape. Consider drawing additional elements and analyzing them until you become familiar with the function.

Because there are no dimensions or text, the LOGICAL option will not produce any results, but you will use ANALYSIS again later. It is possible to obtain far more information about the I beam shape than can be achieved by using ANALYSIS on its own.

Exercise 3: Creating Palettes

If you wish to know the cross sectional area or moment of inertia, the SHAPE function is required. However, SHAPE does not appear on the LPFK palette. This then is an opportunity to create additional palettes.

1. To create a palette for functions not on LPFK, access the Palette Creation screen. Select FILE > KEYBOARD.

2. Click on the Keyboard button to the right of the fixed menu bar at the bottom of the screen. The keyboard area is referred to as the current LPFK because it is laid out like the LPFK keyboard described in the "Basic Hardware" section in Chapter 1.

3. To produce a blank palette, you need to reset by clicking on the RESET button. Note that each of the keys is now blank. A list of available functions appears at the right.

4. To allocate a function to a "key," select it from the list, and then click on the key on which you wish to place the function. The previous illustration shows a palette with all functions not included on the LPFK palette.

5. Allocate the functions to the keys using the method described. When complete, click on the STORE button, and then type in a name for the new palette. The name suggested is *MIX*; henceforth, this palette will be referred to as MIX.

➥ **NOTE:** *If the desired function cannot be seen on the list, scroll up or down using the double arrow located at lower right of the list panel.*

Exercise 3: Creating Palettes

6. To exit the screen, either select another function or click on the EXIT button on the fixed menu if you used method two to access the Palette Creation screen. To switch between palettes, click on the current palette name and select the new one.

To continue with the analysis of the I beam shape, you need to select the SHAPE function. Change to the MIX palette as described above and select SHAPE > UNSPEC.

The I beam shape must be transformed into a closed domain to permit further analysis.

1. To achieve the closed domain, select an element from the I beam geometry. The SEL LN/CRV / / YES:AUTO? message will be displayed in the prompt area. To close the shape, click on the YES button.

2. The SEL LN/CRV / / YES:CLOSE message will now appear; click on the YES button again. Although nothing appears to have changed, in fact a duplicate I beam shape has been created on top of the existing geometry, a closed domain or shape. This can now be analyzed as before using the ANALYSIS function.

3. Select ANALYSIS > NUMERIC. If you now select an element of the I beam, the shape will be selected in preference to the element underneath and the section characteristics will be displayed.

   ➡ **NOTE:** *The use of the SHAPE function shown here is by far the most common. Consider experimenting with other menu options to see how they work.*

4. If you erase an element by selecting it while in the ERASE function, nothing will appear to happen. The explanation is that the shape has been deleted. If you wish to perform additional tasks on the draw elements, you will need to delete or hide the shape because it cannot be used for activities other than analysis or adding patterns.

Chapter 4: Analyzing 2D Graphic Elements

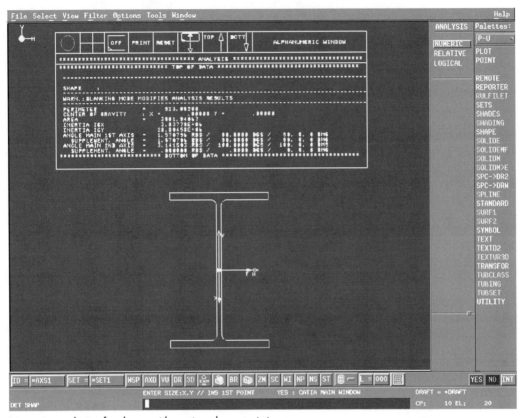

Numeric analysis of a shape with section characteristics.

The PATTERN function in CATIA has been considerably enhanced over the years. Consequently, some SHAPE function uses are no longer needed.

Summary

Topics covered in this chapter include the numeric and relative analysis of 2D elements, the creation of closed domains (shapes), and the analysis of 2D elements and closed domains.

5 Managing 2D Graphic Elements

The purpose of this chapter is to demonstrate the use of CATIA functions for effective management of DRAW elements. An example of DRAW element management would be creating a draw view with several repeated features. With the use of management tools, the feature need only be drawn once. The feature can then be duplicated in as many positions as required, or stored for use in other models as a standard item. Another example would be visualizing only part of a drawing. In this case, you may be able to quickly view discrete parts of the drawing in a single or multiple windows. These examples provide a small sampling of the type of management tasks that you can perform. Appearing below is a list of the most commonly used element management functions.

Element Management Function	Description
TRANSFOR	Moving geometry around the screen.
GROUP, DETAIL, SETS	Grouping elements for use in several positions or to make movements or transformations easier.
LIBRARY	Storing frequently used standard items or groups of elements.
LAYER	Allocate elements to specific areas that can be visualized separately or in groups.
IMAGE	Set up windows and groups of windows (screens) to make effective use of the work area.

58 Chapter 5: Managing 2D Graphic Elements

All functions enumerated in the previous table and respective options are introduced in this chapter. Discussion in subsequent chapters frequently returns to the same functions.

TRANSFOR in DRAW Mode

The TRANSFOR (transformation) function in DRAW mode is used to apply transformations to elements.

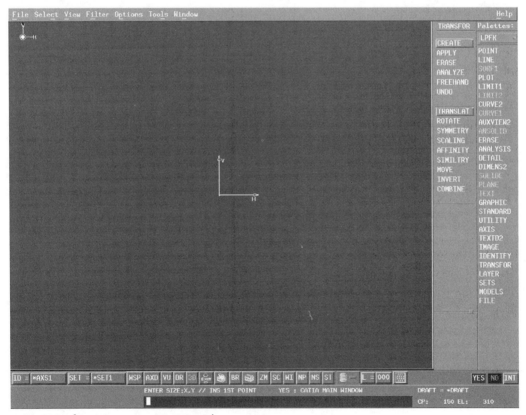

TRANSFOR function menu in DRAW mode.

Option	Description
CREATE | TRANSLATE	Create a translation.
CREATE | ROTATE	Create a rotation.
CREATE | SYMMETRY | LINE | NORMAL or OBLIQUE	Create symmetry about a line.
CREATE | SYMMETRY | POINT	Create symmetry about a point.
CREATE | SCALING	Create a scaling.
CREATE | AFFINITY | NORMAL or OBLIQUE	Create an affinity, a scaling in one direction.
CREATE | SIMILRTY	Define a similarity. (A similarity is the combination of a rotation, translation, and scaling.)
CREATE | MOVE	Define a move. (A move is a combination of a rotation and translation.)
CREATE | INVERT	Create an inverted transformation; available only on stored transformations.
CREATE | COMBINE	Combine transformations.
APPLY | REPLACE | ELEMENT or SET or FAMILY	Apply a transformation in replace mode.
APPLY | DUPLICATE | ELEMENT or SET or FAMILY | STANDARD or SAME	Apply a transformation in duplicate mode.
UNDO	Define number of undo steps available.
ERASE	Erase a transformation.
ANALYZE	Analyze a transformation.
FREEHAND | TRANSLATE or ROTATE | DEFAULT or PIE CHRT or SCALING	Perform freehand movements.

→ **NOTE:** *As seen in the previous table, many TRANSFOR options are available. However, only the most commonly used options (TRANSLATE, ROTATE, and SYMMETRY) are covered in this chapter.*

GROUP in DRAW Mode

The GROUP function in DRAW mode is used to group elements of a model for the same processing.

Chapter 5: Managing 2D Graphic Elements

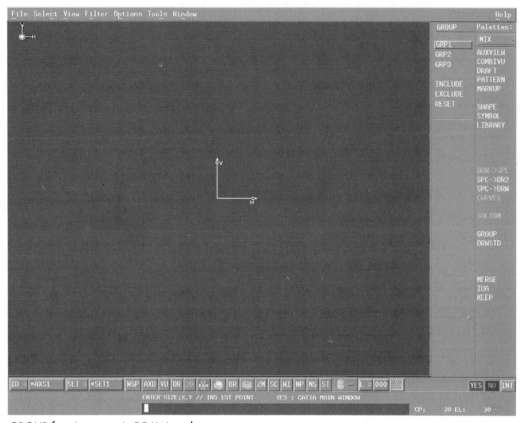

GROUP function menu in DRAW mode.

Option	Description
INCLUDE	Include elements in a selected group.
EXCLUDE	Exclude elements from the selected group.
RESET	Exclude all elements from the selected group.
INCLUDE	TRAP	PART IN or FULL IN or PART OUT or FULL OUT	To define a group using a trap.

Include or exclude elements in the selected group by the following methods of selection.

- INCLUDE or EXCLUDE | CURRENT or W.SPACE | ELEMENT
- INCLUDE or EXCLUDE | CURRENT or W.SPACE | TYPE
- INCLUDE or EXCLUDE | CURRENT or W.SPACE | LAYER
- INCLUDE or EXCLUDE | CURRENT or W.SPACE | LAYER
- INCLUDE or EXCLUDE | CURRENT or W.SPACE | GROUP
- INCLUDE or EXCLUDE | CURRENT or W.SPACE | STRING
- INCLUDE or EXCLUDE | CURRENT or W.SPACE | FAMILY
- INCLUDE or EXCLUDE | CURRENT or W.SPACE | ATTRIBUT
- INCLUDE or EXCLUDE | CURRENT or W.SPACE | DRW ELEM

DETAIL in DRAW Mode

The DETAIL function in DRAW mode is used to place elements in an alternative workspace which can be used on the master workspace as a ditto (or a copy) of its elements, to perform transformations and analysis on dittos, and to switch between workspaces. Options are described below.

Chapter 5: Managing 2D Graphic Elements

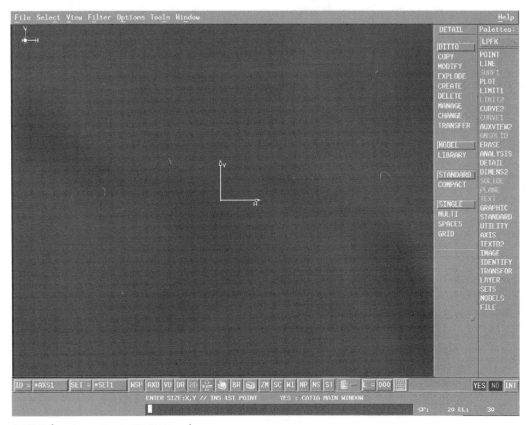

DETAIL function menu in DRAW mode.

Option	Description
DITTO | MODEL | STANDARD or COMPACT | SINGLE or MULTI or SPACES or GRID	Create a ditto from a detail existing in the model.
DITTO | LIBRARY | STANDARD or COMPACT | SINGLE or MULTI or SPACES or GRID	Create a ditto from a detail stored in a library.
COPY | MODEL	Copy the elements of a detail in a model to a detail or view.
COPY | LIBRARY	As above but for a library detail.
MODIFY | TRANSLATE	Translate a ditto or a copy.
MODIFY | ROTATE	Rotate a ditto or a copy.
MODIFY | SCALE	Modify the scale of a ditto or a copy.

MODIFY | SYMMETRY	Create a symmetry of a detail or copy.
MODIFY | FLIPX or FLIPY	Perform symmetry about the current axis of a ditto.
EXPLODE	Transform a ditto into a geometric copy.
CREATE	Create a new detail.
DELETE | UNSED or USED | DIRECT or VISUALIZATION | DETAIL or DITTOS	Delete a detail.
MANAGE | ANALYZE | DETAIL or W.SPACE	Analyze a detail or a workspace.
MANAGE | UPDATE	Update a library detail and its family with respect to the current library.
MANAGE | REPLACE | W.SPACE or CURRENT	Replace a ditto by another ditto.
MANAGE | VERIFY | CURRENT or W>SPACE	Verify the nature of dittos.
MANAGE | LAYER | CURRENT or W>SPACE	Change standard dittos into compact dittos or vice versa.
MANAGE | RENAME	Change the name or comment of a detail.
MANAGE | DROP	Delete the link between a detail stored in a library and the current model.
CHANGE	Change the current workspace.
TRANSFER | REPLACE or DUPLICAT | ABSOLUTE or RELATIVE	Transfer elements from one workspace to another.

☛ **NOTE:** *DETAIL function options are numerous. Only the most frequently used options are discussed in this chapter.*

LIBRARY in DRAW Mode

The LIBRARY function in DRAW mode is used to define the current library, write objects to a library, read objects from a library, update objects stored in a library, and delete objects stored in a library. Options are described below.

Chapter 5: Managing 2D Graphic Elements

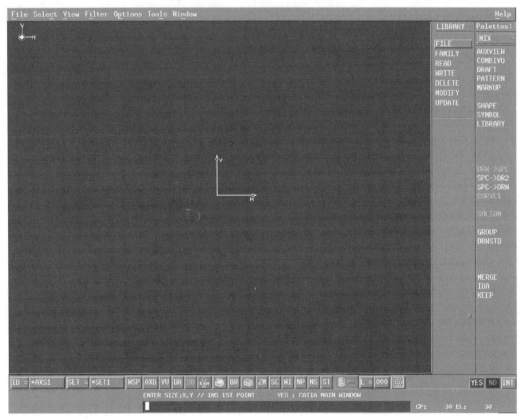

LIBRARY function menu in DRAW mode.

Option	Description
FILE	Select the library file where the families and objects to be handled are defined.
FAMILY	Select the family where the objects to be handled are defined.
READ	Find a library stored object to be incorporated into the model.
WRITE	Write a model into a library.
DELETE	Delete an object from a library.
MODIFY	Modify an object's keywords.
UPDATE	Update a library stored object according to model modifications.

SETS in DRAW Mode

The SETS function in DRAW mode is used to manage sets of elements by creating, deleting, changing, or linking sets. To change the current set from pick to no-pick or vice versa, use CHANGE | NO PICK or PICK. To create a set, use CREATE, and to delete a set, DELETE.

Transfer elements from one set into the current one with TRANSFER. Copy one or more elements into the current set with COPY | STANDARD or SAME. Finally, to link a set with the current set, use LINK.

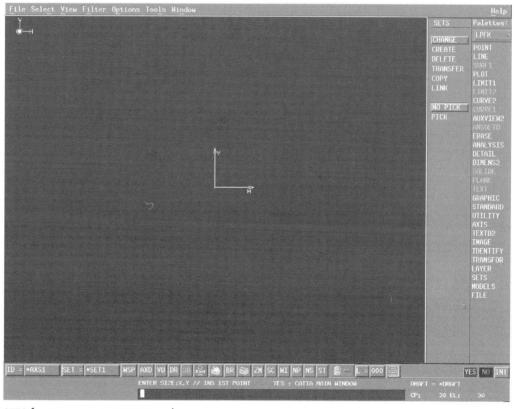

SETS function menu in DRAW mode.

LAYER in DRAW Mode

The LAYER function in DRAW mode is used to apply a layer filter to the screen, view, or ditto; create and modify filters; change the current layer; transfer elements between layers; and analyze layers and filters. Options are described in the next table.

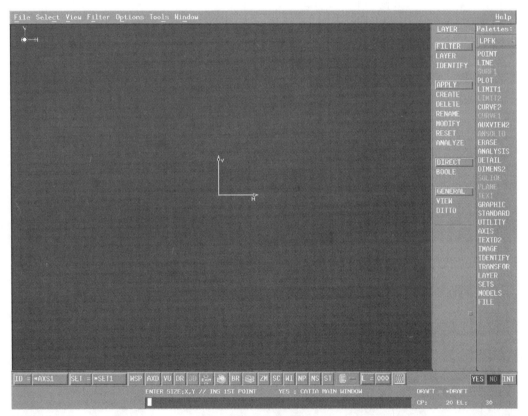

LAYER function menu.

Option	Description
APPLY	Apply a layer filter.
FILTER | APPLY | DIRECT	Apply a layer filter selected from the list of existing filters.
FILTER | APPIY | DIRECT | GENERAL	Apply a layer filter selected from the list to the screen.

FILTER | APPLY | DIRECT | VIIEW	Apply a layer filter selected from the list to a DRAW view.
FILTER | APPLY | DIRECT | DITTO	Apply a layer filter selected from the list to a ditto.
FILTER | APPLY | BOOLE | GENERAL	Apply a display layer filter created from combinations of other filters to the screen.
FILTER | APPLY | BOOLE | VIEW	Apply a display layer filter created from combinations of other filters to a DRAW view.
FILTER | APPLY | BOOLE | DITTO	Apply a display layer filter created from combinations of other filters to a ditto.
FILTER | CREATE	Create a filter.
FILTER | CREATE | DIRECT | VIEW or DITTO or GENERAL	Create a display layer filter from the list to a DRAW view, ditto, or screen.
FILTER | CREATE | BOOLE	Create a layer filter from combinations of other filters.
FILTER | DELETE	Delete filters.
FILTER | RENAME	Rename a filter.
FILTER | MODIFY | CUR > FILT | GENERAL or VIEW or DITTO or OTHER	Modify the status of a layer filter applied to a DRAW view, ditto, solid, or screen.
FILTER | RESET | W.SPACE or CURRENT	Reset filters.
FILTER | ANALYZE	Analyze filters.
LAYER | CHANGE	Change layers.
LAYER | TRANSFER | W.SPACE or CURRENT	Transfer elements from one layer to another.
LAYER | VERIFY | NO PICK or PICK	Display the elements on the current layer.
LAYER | ANALYZE	Analyze the layers currently used.
IDENTIFY	Create or modify the layer table in the project file.

IMAGE in DRAW and SPACE Modes

The IMAGE function in DRAW and SPACE modes is used to manage the display of model elements and screen windows. Options are described in the next table.

> **NOTE:** *This function does not have separate DRAW and SPACE mode uses because it is used to handle both DRAW and SPACE windows at the same time.*

Chapter 5: Managing 2D Graphic Elements

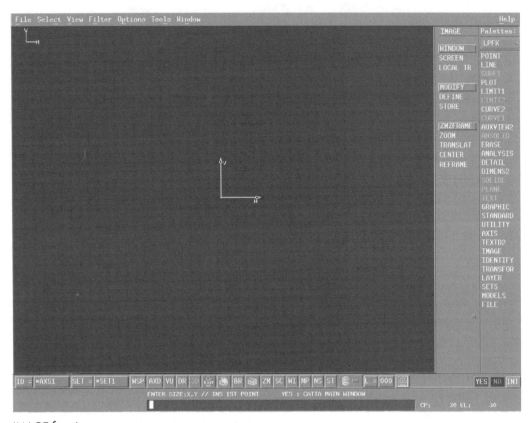

IMAGE function menu.

Option	Description
WINDOW I MODIFY I PLANE	Modify a window display in the screen plane.
WINDOW I MODIFY I PLANE I SCALE or ZOOM	Modify the window scale or centering.
WINDOW I MODIFY I PLANE TRANSLAT	Translate the model or sheet display in a window.
WINDOW I MODIFY I PLANE I CENTER	Move the origin of the current two- or three-axis system to the middle of a single window screen or the middle of a window.
WINDOW I MODIFY I PLANE I REFRAME	Reframe the model display in a window by optimizing its centering and scale.
WINDOW I MODIFY I SPACE I ROT UNSP or ROT VERT or ROT HOR or ROT PLN	Rotate the model about a line of the model or in the screen plane.

Exercise 1: Using TRANSFOR

WINDOW | MODIFY SPACE | EYE MOVE	Move the center of projection a given distance along line of sight.
WINDOW | MODIFY | SPACE | CENTER	Move the line of sight.
WINDOW | MODIFY | SPACE | EYE CNTR	Move the projection center in a direction parallel to the screen.
WINDOW | DEFINE	Define a cylindrical projection.
WINDOW | DEFINE | CONIC	Define a conic projection perspective.
WINDOW | DEFINE | UNSPEC	Define a general unspecified projection.
WINDOW | STORE	Store a window.
WINDOW | RECALL	Recall a stored window.
WINDOW | RENAME	Modify the identifier of a stored window.
WINDOW | DELETE	Delete stored windows.
SCREEN | MODIFY	Modify all active windows simultaneously.
SCREEN | MODIFY | ZM/FRAME	Modify the screen scale or centering.
SCREEN | MODIFY | ZOOM	Modify the screen scale or centering.
SCREEN | TRANSLATE	Translate a screen.
SCREEN | MODIFY | SEPARATE	Translate the screen separation lines.
SCREEN | DEFINE	Define one or more windows within the working area of the screen.
SCREEN | STORE	Store a screen.
SCREEN | RECALL	Recall a stored screen.
SCREEN | RENAME	Modify the identifier of a stored screen.
SCREEN | DELETE	Delete a screen.
LOCAL TR	Activate or deactivate local transformations.

Exercise 1: Using TRANSFOR

First, read the I beam model you saved at the end of Chapter 3 with FILE > READ. To demonstrate the TRANSFOR function, let's backtrack a few steps. Using the ERASE and LIMIT1 functions, modify the model geometry so that it resembles the next illustration. On the other hand, you could start from scratch by using the functions discussed in the first three chapters to recreate the geometry in the illustration.

A quarter of the I beam geometry.

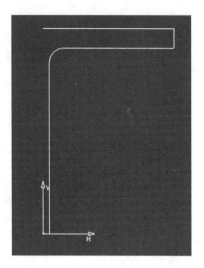

Before proceeding, save the model with a different name (e.g., *I BEAM ONE QUARTER*). You may have noticed that the I beam has symmetry about two center lines. Consequently, it is necessary to draw one quarter of the elements. The remaining three quarters can be produced using the TRANSFOR function.

1. Select TRANSFOR > CREATE | SYMMETRY | LINE | NORMAL.

2. Select the lines of symmetry, beginning with the vertical center line. A red double arrow appears close to the selected line and the SYMMETRY DEFINED message displays. At this point the defined transformation can be stored by typing a file name and pressing the <Enter> key. Alternatively, you can proceed straight to the APPLY option.

Exercise 1: Using TRANSFOR 71

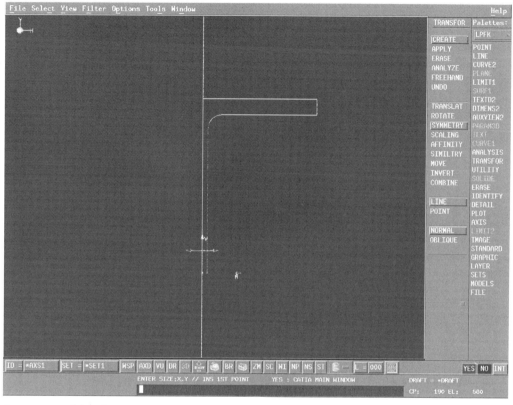

One quarter of the I beam geometry.

3. Select TRANSFOR > APPLY | DUPLICATE | SAME.

⊷ **NOTE:** *The DUPLICATE and SAME selections are made in the window that appears when the APPLY option is selected.*

Transformation Apply window.

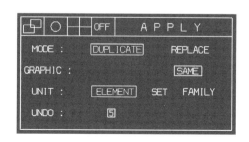

Chapter 5: Managing 2D Graphic Elements

4. Select the elements that you wish to transform, that is, the elements comprising the one quarter geometry of the I beam. As you select the elements, they are duplicated by symmetry around the selected vertical line.

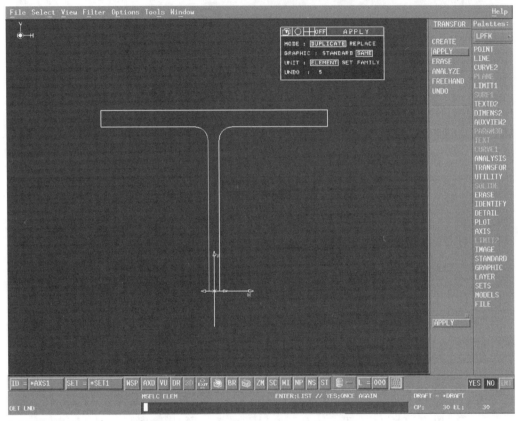

Selecting elements for transformation.

5. Define the symmetry around the horizontal center line. Follow steps 1 to 3 using the horizontal center line. When you apply the transformation, you will need to select the elements that define a half of the I beam geometry.

The original I beam should be recreated based on a mere quarter of the geometry. There is no need to save the drawing because it was already saved in Chapter 3. It may have occurred to you that the above procedure would have been easier if the elements for transformation were selected as a group. The next exercise explores how to group elements.

Exercise 2: Using GROUP

Read the *I BEAM ONE QUARTER* model saved in the previous exercise with the FILE > READ command. In this exercise, the complete I beam is recreated, but this time by grouping the elements on which the symmetry will be performed. The first method will use the GROUP function.

1. Select GROUP > GRP_1 | INCLUDE | CURRENT | ELEMENT. Now you need to select the elements to be included in group 1. Select each of the individual elements that comprise the one quarter geometry. As you select the elements, they will temporarily disappear from view.

2. When you have selected all the elements, check what has been included in *GRP_1* (group 1) by selecting the EXCLUDE option. This option shows you the elements that can be excluded from the group.

3. You are now ready to execute the symmetry in precisely the same way as in Exercise 1. Select TRANSFOR > CREATE | SYMMETRY and then select the line of symmetry.

4. Select TRANSFOR > APPLY | DUPLICATE | SAME. When you apply the transformation, type *GRP1* in the input information area and press the <Enter> key. The entire I beam quarter is transformed by symmetry around the selected line.

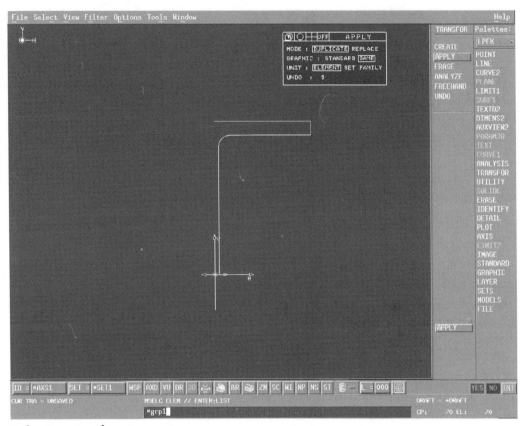

Performing a transformation on a group.

> **NOTE:** *Applying a transformation to a group of elements is called "multi-selection." See the multi-select options listed in Appendix B.*

The next phase of the process is to include the newly created quarter of the model in *GRP_1*.

1. Select GROUP > GRP_1 | INCLUDE | CURRENT | ELEMENT. The first quarter disappears because these elements are already included in *GRP_1*. As you select the elements from the second quarter, they too will disappear as they are included in *GRP_1*.

2. Execute the symmetry about the horizontal center line. Select TRANSFOR > CREATE | SYMMETRY and then select the line of symmetry. Select TRANSFOR > APPLY | DUPLICATE | SAME. When you apply the transformation, type *GRP1* in the input information area and press the <Enter> key. This time the top half of the I beam will be transformed around the selected line, and the I beam is once again complete.

The time savings resulting from grouping are not readily apparent in this simple exercise. Imagine that you are working with 105 elements in the quarter shape rather than a mere five.

Exercise 3: More Transformation Options

Read the *I BEAM ONE QUARTER* model with the FILE > READ command. Repeat the steps in the previous exercise until you reach the stage of defining the transformation, and then take the following steps.

   1. Select TRANSFOR > CREATE | ROTATE. The message ROTN CENTER: SEL PTD // SEL LND1 will be displayed. CATIA is asking you to select a rotation center.

   2. Because the rotation center will be the origin in this exercise, input a comma (,) and press <Enter>. A temporary point will appear with a red circle around it.

   3. The message ANGLE: SEL START PTD // SEL LND1 SEL CIRD // KEY ROTATION ANGLE will be displayed. CATIA is asking you to define the rotation.

   4. Type *180* <Enter>, and a graphic showing the defined rotation and a message will be displayed. If you want to store this transformation, input a file name and press <Enter>. Type *ROT180* <Enter>. The 180 degree rotation is stored.

To view the list of stored transformations, Select TRANSFOR > APPLY and then press <Enter>. You will see two items in the list: IDENTITY ROT180. The ROT180 transformation will be highlighted which means that it is the current transformation, and therefore cannot be reselected. Reselect APPLY from the menu.

You should now apply the transformation by selecting individual elements or using GROUP. Do not forget to select the duplicate option. The one quarter geometry will be rotated to the position shown below.

Half complete I beam after 180 degree rotation.

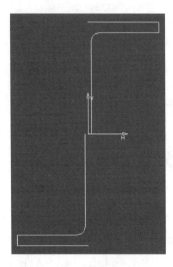

Exercise 4: Using the DETAIL Function

Read the side plate model saved in Chapter 3 with the FILE > READ command. (To create this shape from scratch, you could draw only half the elements and then use the TRANSFOR function's SYMMETRY option. Consider creating the side plate from scratch in this way in a practice session.) Take the following steps to experiment with the DETAIL function.

1. Erase all but one of the holes and respective center lines. Leave the points on the centers of the erased holes. Your screen should resemble the following illustration.

Exercise 4: Using the DETAIL Function

Side plate containing a single hole.

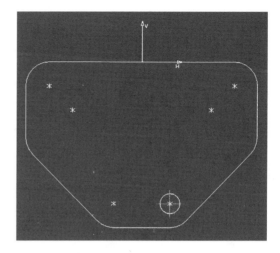

2. The next step is to transfer the elements of the hole and center lines into a detail workspace using the automatic creation facility in the DETAIL function's transfer option. Select DETAIL > TRANSFER | REPLACE | RELATIVE. Click on the YES button three times; first, to continue, next to confirm no translation, and finally to confirm rotation.

3. Key in *Sel* to the input information area and press <Enter> to begin the multi-select process.

4. Select the circle element (the hole) and then its center lines. The elements will be highlighted as you select them.

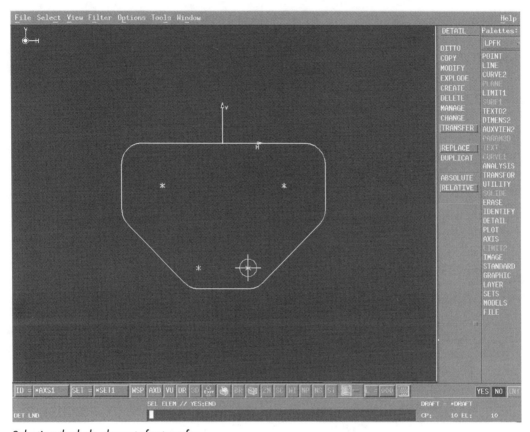

Selecting the hole elements for transfer.

5. Click on the YES button to end the multi-select. Click on the YES button again to transfer the elements to the detail workspace. As the elements transfer, they disappear from the screen.

6. Select DETAIL > DITTO | MODEL | COMPACT | MULTI. The name of the last detail used or created will appear in the data entry panel. Press on <Enter> to select this detail. Select one of the hole center points on the model. The ditto of the side plate hole detail will be created on the selected point. Individually select the other three hole center points; dittos are created on each.

Exercise 4: Using the DETAIL Function 79

Completed side plate.

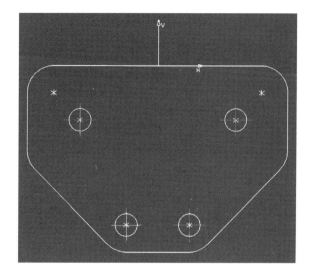

To demonstrate how quickly you can produce additional dittos of the holes using mouse button 2, indicate various points around the screen. A ditto will be created on each point. To tidy up, erase the unwanted holes.

An alternative approach to that presented in the previous steps would be to create a single ditto of the hole, and then move it to other positions by defining a translation. This approach is described in the following steps.

1. Select TRANSFOR > CREATE | TRANSLATE. Select the center point of the hole ditto and then select the point on the center of the next hole. A red arrow will be displayed as well as the TRANSLATION DEFINED message.

2. Select TRANSFOR > APPLY | DUPLICATE | ELEMENT | SAME. When you select the ditto of the hole, it will be copied to the selected position. This procedure can be followed for the other two holes.

3. Select DETAIL > DITTO | MODEL | COMPACT | SINGLE. Create a single ditto of the hole on one of the center points.

4. Select DETAIL > MODIFY | TRANSLATE | DUPLICATE. Select the ditto to be duplicated and then select one of the other center points. Repeat for the other holes.

5. Save the side plate model containing dittos with the following name: *SIDE PLATE WITH DITTOS*.

In CATIA, there are typically many methods available to achieve a particular result. As you gain experience, you will make choices about which method is appropriate for each task.

LIBRARY Function

The use of LIBRARY will vary significantly according to your project setup. The principal purpose of this function is to store frequently used or standard items (library details). Diverse attributes can be assigned to library details, such as drawing number, material, and supplier.

As long as the link is maintained between a library ditto and its respective library, the ditto can be updated to reflect the current status of details. Alternatively, library details can be updated to reflect changes to standard items. One common use of the library function is to store drawing sheets. (See Chapter 7.) To determine whether a library facility is available and to obtain information on library organization, consult your system administrator.

Exercise 5: Using SETS

If you still have the side plate model with dittos on the screen, then proceed. If not, read the model before you continue with the FILE > READ command.

1. Using the TRANSFOR function, create a translation to copy the geometry 350mm (13.79") to the right, either by creating points and using them or by selecting a horizontal line and entering a distance of 350mm. Select TRANSFOR > CREATE | TRANSLATE.

2. Now that the translation is defined, you can apply it. Select TRANSFOR > APPLY, and from the APPLY window select DUPLICATE | SAME | SET.

3. Select an element. All elements will be translated and duplicated 350mm. In the process, a new set has been created.

Exercise 6: Using LAYER 81

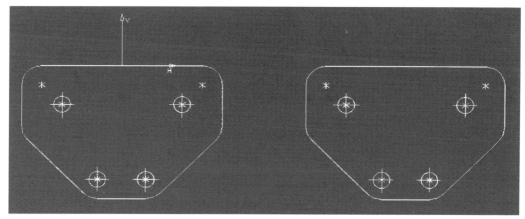

Duplicate side plates.

    4.   To examine sets in more detail, select the SETS function from the LPFK palette. The duplicate side plate appears to be dimmed, indicating that these elements are in the current set. Select an element from the original geometry. You have now returned to the original set. If you repeat the last two steps while watching the SET button at the left end of the fixed menu, you will see the set number change.

    NOTE: *The SET button can be used to change sets while using another function.*

    5.   Save the model using the name *SIDE PLATE DUPLICATE*.

Several of the SETS function's other options (e.g., CREATE, DELETE, and TRANSFER) will be used in subsequent chapters.

Exercise 6: Using LAYER

Read the *SIDE PLATE DUPLICATE* model saved in the last exercise. To the present, all drawings have been executed on layer 0. To the right of the L= button at the right of the fixed menu, *000* appears on another button. The latter button indicates the current layer, in this case, layer *000*.

In the following steps, you will transfer elements to a different layer and then create filters to visualize the layers individually or in groups. Filters can generally be applied to all items on the screen or selectively by view or ditto.

✓ **TIP:** *Imagine that layers are transparent drawing sheets containing various elements. The transparencies can be viewed individually, so that only what is drawn on a single sheet is seen. On the other hand, several transparencies can be stacked on top of each other, thereby enabling a simultaneous view of all elements drawn on all transparencies. The latter set of transparencies would be equivalent to a filter.*

To demonstrate how layers and filters work, take the following steps.

1. Select LAYER > TRANSFER | W.SPACE. Type 99 <Enter>. Type *set <Enter> to employ the multi-select facility. Select any element. All elements in the set will be highlighted.

2. Click on the YES button to confirm the multi-select, and then click on the YES button to end the task. You have now transferred some of the elements to layer 99.

3. Select LAYER > FILTER | CREATE. Select NONE from the Filters window at the top right of the work area, and verify that the GENERAL item is highlighted in the FILTER MODE window at the bottom left of the work area. From the displayed list of layers, select layer 099. (Click on the double arrow at the bottom right of the panel to scroll to the next page.)

4. Select *LAYER 99*; the item will change color from green to white. Click on the YES button to create the filter. Input a name for the filter such as *F99* and press <Enter>. As seen in the Filters window, the new filter is now in the list.

Exercise 6: Using LAYER

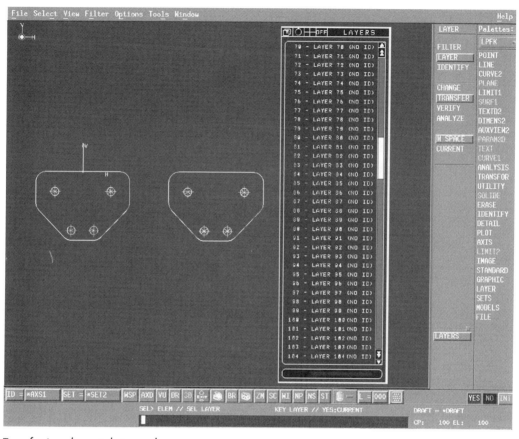

Transferring elements between layers.

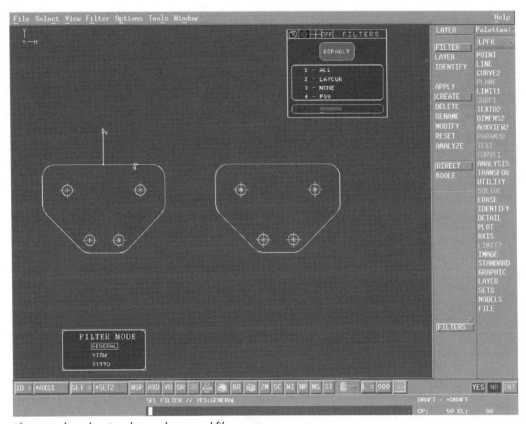

Filters window showing the newly created filter.

5. To view the effect of using the newly created filter, you need to change the current layer. (The current layer is always visualized.) Click on the L = button on the fixed menu bar and then click on the YES button to move forward to layer *001*. Now click on the blue EXIT button in the center of the fixed menu bar.

6. Select LAYER > FILTER | APPLY | DIRECT | GENERAL, and then select the *F99* filter. Elements on layer *000* are visible.

7. Apply the *ALL* filter; all elements are visible once again. Save the model as *CHAPTER 5 EXERCISE 6*.

Additional options for the LAYER function will be explored in subsequent chapters.

Exercise 7: Using IMAGE

This exercise provides an introduction to the IMAGE function, which allows you to create windows and groups of windows called "screens." Assume that you wish to enlarge one of the holes in the side plate to perform closeup work.

1. Using mouse button 3 and dial 3 (if you have dials) or the <Shift> and cursor keys, zoom in to one of the side plate holes so that it fills the screen. To store this view as a window, select IMAGE > WINDOW | STORE and give the window a name by inputting *w1* and pressing <Enter>.

   NOTE: *Zooming and panning while working with the IMAGE function can be performed only if the LOCAL TR (local transformations) option is selected from the menu.*

2. Zoom, pan, and drag so that one of the side plates fills the screen, and then store the window as *w2*. Using the same method, create a window that shows both side plates and store it as *w3*.

3. To view the stored windows, click on the WI button on the fixed menu bar, and then click on the YES button repeatedly to scroll through the stored windows.

4. Select IMAGE > SCREEN | DEFINE. The SCREEN DEFINITION window appears.

Chapter 5: Managing 2D Graphic Elements

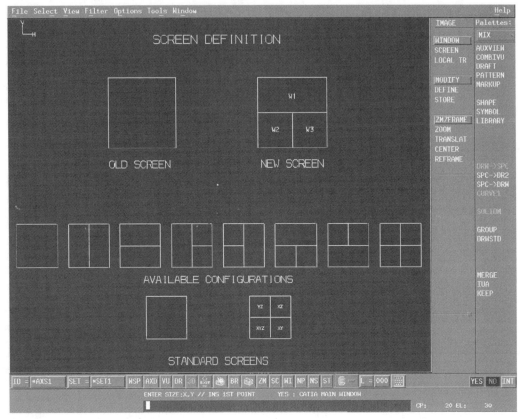

SCREEN DEFINITION window.

5. From the row of eight screen configurations, select the option shown in the next illustration.

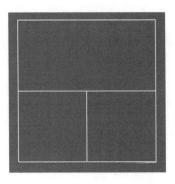

Three-window screen configuration.

Exercise 7: Using IMAGE

6. The selected configuration will appear in the NEW SCREEN location with one of the window positions highlighted. A message asking you to select a window appears. Click on the LIST button to view the list of stored windows and from the list select *w3*.

7. Repeat steps 4 through 6 to place the other two windows in the new screen and the click on the YES button to confirm creation. The screen should now resemble the next illustration showing three windows of the model. To make any of the windows live, click on the axis icon in its respective corner.

✓ **TIP:** *Do not forget that in the IMAGE function, local transformations are possible only if the LOCAL TR button has been selected, or if you change to a different function.*

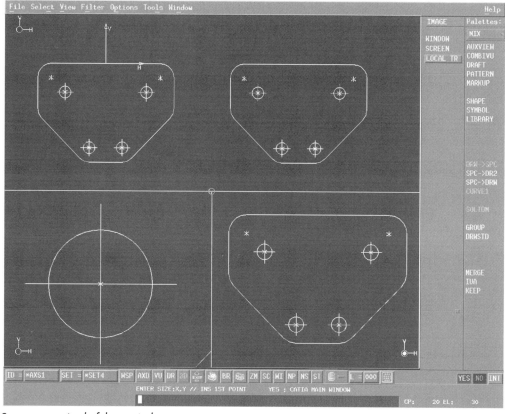

Screen comprised of three windows.

This exercise focused on 2D windows only. In subsequent chapters you will see how a mixture of 2D and 3D windows can be stored together in a single window.

Summary

Topics covered in this chapter included an introduction to using libraries, and how to define and apply a transformation; group items together for common processing; create details and using dittos; group items in sets; place items on different layers; make filters; and create windows and screens.

6 Creating Multiple Views

Once you are familiar with generating, modifying, analyzing, and managing elements, you need to know how to create more than one view in a drawing. Multiple views are orthographic views. Unlike drawing on paper, or many other CAD systems, CATIA creates views that are linked to each other to enable geometry from one view to be used to generate geometry in another view.

The purpose of this chapter is to explore the following functions that can be selected from the palette menus.

- AUXVIEW (define and lay out views)
- COMBIVU (exchange geometrical information between views)
- AXIS (create additional axis systems within a view)
- DRAFT (create additional drafts in order to generate different drawings from the same views)

In this chapter you will use exercises from previous chapters to create additional orthographic views to generate complete drawings.

AUXVIEW

The AUXVIEW function is available only in DRAW mode. Options are described in the following table.

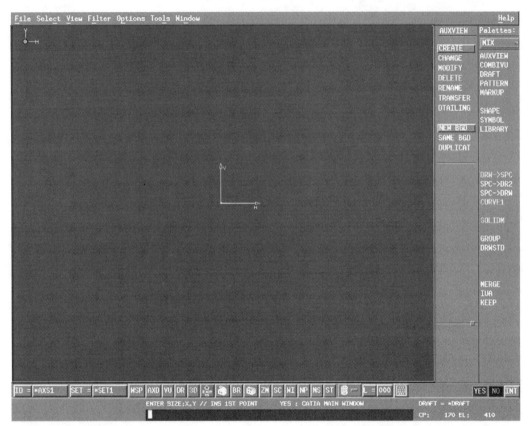

AUXVIEW function menu.

Option	Description
CREATE	Create views.
CHANGE	Swap from one view to another.
MODIFY	Modify, change scale, and translate views.
DELETE	Delete views that are no longer required.
RENAME	Change view names.
TRANSFER	Move elements from one view into another.
DTAILING	Show and no show text, dimensions, or patterns in the view.

COMBIVU

The COMBIVU function is available only in DRAW mode. Options are described below.

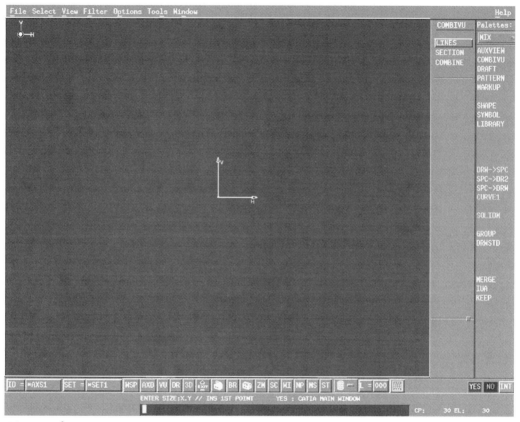

COMBIVU function menu.

Option	Description
LINES	Create lines in the current view from elements selected in another view.
SECTION	Create cross sections of cylinders in current view.
COMBINE	Create elements in current view by combining plane elements from other views.

DRAFT

The DRAFT function is available only in DRAW mode. Options are described in the next table.

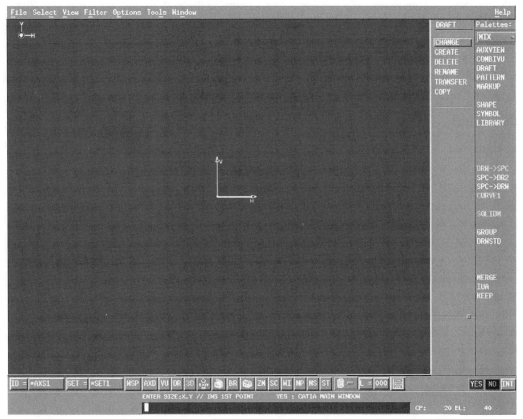

DRAFT function menu.

Option	Description
CREATE	Create a new draft.
CHANGE	Swap from one draft to another.
DELETE	Delete drafts no longer required.
RENAME	Change the name of drafts.
TRANSFER	Move elements from one draft to another.
COPY	Duplicate an existing draft.

AXIS

AXIS function options are described below.

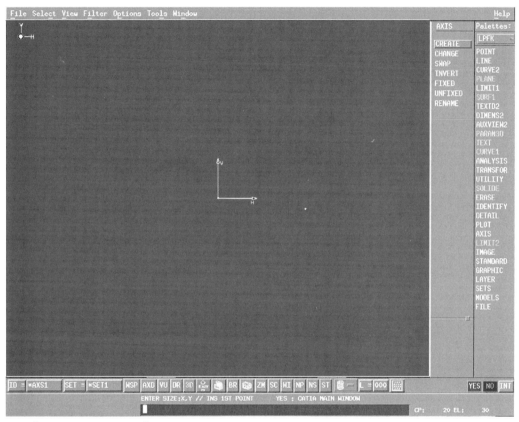

AXIS function in DRAW mode.

Option	Description
CREATE	Create a new axis system.
CHANGE	Swap from one axis system to another.
SWAP	Swap the horizontal and vertical axes of an axis system.
INVERT	Invert the axis of an axis system.
FIXED	Change an unfixed axis system into a fixed axis system.
FIXED	Change a fixed axis system into an unfixed axis system.
RENAME	Change the name of an axis system.

Exercises in this chapter are designed to provide you with experience in using some of the most frequently used options under the AUXVIEW, COMBIVU, DRAFT, and AXIS functions. The first exercise provides an introduction to AUX-VIEW and COMBIVU. Exercises 2 through 4 present more complex demonstrations of the four functions.

Exercise 1: Using AUXVIEW and COMBIVU

The object of this exercise is to create three orthographic views and an isometric view of the equal angle section shown in the next illustration.

Equal angle section with isometric view.

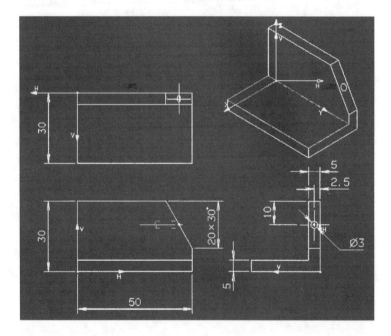

Creating the Front View

The first view to create in this exercise is the front elevation. The steps to create this view follow. Functions and options required to create most of the view have been discussed in previous chapters. Consequently, only the required steps are outlined.

Exercise 1: Using AUXVIEW and COMBIVU

1. Create the two vertical lines using LINE > VERTICAL | UNLIM. Create the three horizontal lines using LINE > HORZONT | UNLIM.

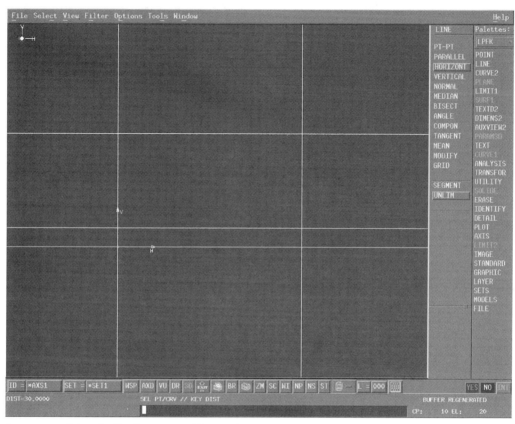

Screen containing the five lines created in step 1.

2. Relimit the lines using LIMIT1 > RELIMIT | TRIM ALL to create the corners. Relimit the lines using LIMIT1 > MACHINE | TRIM ALL to create the chamfer. Remember that because the angle is not 45 degrees (the default), you will have to enter *30* as the required angle.

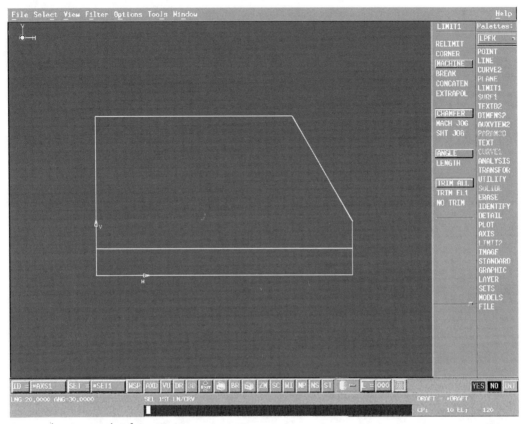

Screen showing results of step 2.

3. The next stage is to create the hidden detail for the hole in the chamfer. To create the center line of the hole, use LINE > PARALLEL | UNLIM. Select the top horizontal line and then indicate anywhere below the line. Key in *10* to the input information area and press <Enter>.

4. Relimit the line using LIMIT1 > RELIMIT | TRIM ALL. Because there are no other elements to trim the line to, select the line and indicate a position where you would like the line to end. When one end of the line has been limited, repeat the process to limit the other end. Do not change the graphical representation of the line; it will be changed later with a GRAPHIC function option.

5. You now need to create the lines for the hole. Select LINE > PARALLEL | UNLIM and create the two horizontal lines at 1.5 on both sides of the center line.

Exercise 1: Using AUXVIEW and COMBIVU

6. The line for hole depth should now be created, but there are no lines available from which to create a parallel line. Consequently, you will create a point on the intersection between the angled line and the hole center line. Select POINT > PROJ INT | SINGLE LIM OFF. Select the two lines to locate the point on the intersection between the angled line and the center line.

7. From this point you can create a vertical line using LINE > VERTICAL | UNLIM. Once you have created this vertical line you can create a parallel line at a distance of *10*.

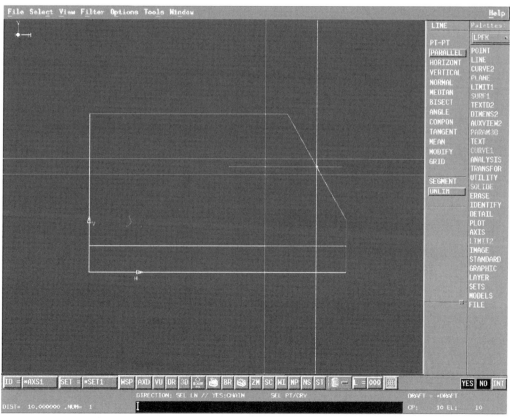

Creating a parallel line at distance of 10.

8. The vertical line through the point, as well as the point, can now be erased using the ERASE function because they are no longer required.

Chapter 6: Creating Multiple Views

Alternatively, the two elements could be hidden using ERASE > NO SHOW. If you hide elements, you can use them again later.

9. The three lines will now be relimited. The two corners at the left end of the hole can be relimited using LIMIT1 > RELIMIT | TRIM ALL in the same manner as above.

10. The two horizontal lines should now be relimited to the angled line. For these lines you will use LIMIT1 > RELIMIT | TRIM EL1 because you want to trim the horizontal lines only and not the angled line. Select LIMIT1 > RELIMIT | TRIM EL1, the top horizontal line, and then the angled line. Note that only the first selected line has been relimited. TRIM EL1 means trim the first selected element only; if you had selected the angled line first, the angled line would have been relimited.

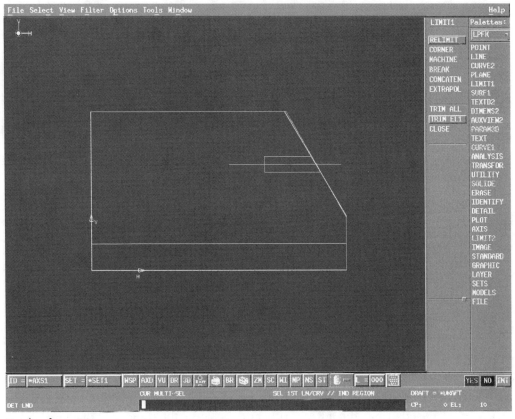

Results of step 10.

Exercise 1: Using AUXVIEW and COMBIVU 99

At this point, only the angled lines for the end of the hole remain to be created for the targeted view. Once again, you will use a new LINE option.

1. Select LINE > COMPON | SEGMENT | ONE LIM. Select the short vertical line that forms the bottom of the hole near its intersection with the top horizontal line. A vector line at an angle from the end of the selected line is highlighted. A point at the end of the selected line is highlighted.

2. In the message area, CATIA asks for an angle. Type *60* in the input information area and press <Enter>. CATIA now asks you to define the limit of the line to be created. Select the center line and the angled line that forms the bottom of the hole.

3. To create the other short angled line, follows steps 1 and 2 and use an angle of *-60*. If you key in *60* instead of *-60*, you can select the highlighted vector line to invert the vector.

Your drawing should now resemble the following illustration, excepting the line designations which will be used to help explain the creation of the next two views. If your view does not resemble the illustration and you have experienced a few problems creating the drawing, refer to previous chapters to check on the functions and options used thus far in this exercise.

Finished front view.

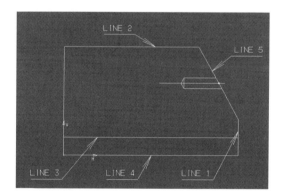

Side View

The next phase of this exercise is creation of the side elevation. If this view was being created on a drawing board, you would simply start drawing the new view by projecting the horizontal lines across from the front elevation. In CATIA, the same basic principle applies, but the first step is to create a new view.

1. Select AUXVIEW > CREATE | NEW BGD (Background). This function and respective options will allow you to create a new view. At this point, you need to define the plane for the new view. In this case, the plane is the right hand vertical line on the front elevation view.

2. Select line 1 as indicated in the previous illustration. INDICATE POSITION YES:SAME ORIGIN appears in the message area. To ensure that the views use the same origin system, and are properly related to one another, you must use the YES option here. Click on the YES button.

3. You are now asked to key in *1* (third angle projection, American view convention) or *2* (first angle projection, European convention). The angle projection used in this exercise is the third, so key in *1* to the input information area. (In subsequent exercises, keying in a *1* is not necessary because the last used convention is active by default.)

4. At this point, you must define the location of the new view. With the use of mouse button 2 indicate the position at the right of the front elevation.

5. CATIA now asks you to name the new view. Key in *SIDE* to the input information area. A new axis system appears. Note that the vertical and horizontal axes have been inverted. When you require a line to be vertical to the screen in the new view you will need to select the horizontal option in the LINE function and vice versa for a line horizontal to the screen.

Exercise 1: Using AUXVIEW and COMBIVU

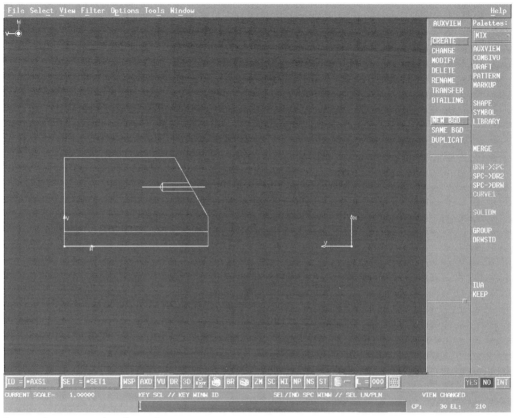

Screen showing results of step 5.

Now that the new view has been created, you need to create geometry.

1. Create the three vertical lines with dimensions available in the target drawing. In this case, use 5 and 30 as seen in the target drawing illustration. Use LINE > HORIZONT | UNLIM to create the three lines vertical to the screen in the side view, but horizontal to the system axis. (Recall the discussion of axis system inversion.)

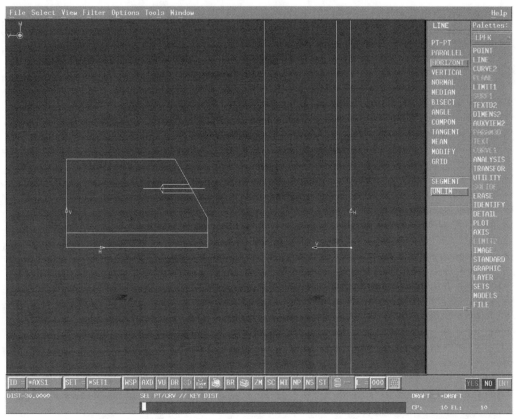

Screen showing results of step 1.

2. To create the other four lines in the view, horizontal to the screen but vertical to the axis, you could use the dimensions available from the drawing. However, entering information always runs the risk of error. To eliminate as many errors as possible, use the information available from the first view (front elevation) and the COMBIVU function. Select COMBIVU > LINES.

If you now select one of the lines in the front view (e.g., line 2), CATIA would create an unlimited line in the new side view. Rather than using this line, you will attempt to keep the view tidy by using the other option, that is, to limit the length of projected lines to existing geometry. If you have created unlimited lines, erase them.

Exercise 1: Using AUXVIEW and COMBIVU

3. Select COMBIVU > LINES. Instead of selecting lines from the front view first, select the two lines in the side view that are five apart. This procedure limits lines for your first line. Now select line 2 in the front view and the line that is created will be limited by the two lines that are five apart.

4. The next line limited by the two lines is the line on the intersection between lines 5 and 1. To create this line, select line 5. CATIA now asks you if you want lines to be created on the limits of this line or to create a line on an intersection. In this example, you will create the line on an intersection. Select line 1; a new line appears in the side view. Because this is the last of the lines limited by the two lines five apart, click on the YES key to end the selection.

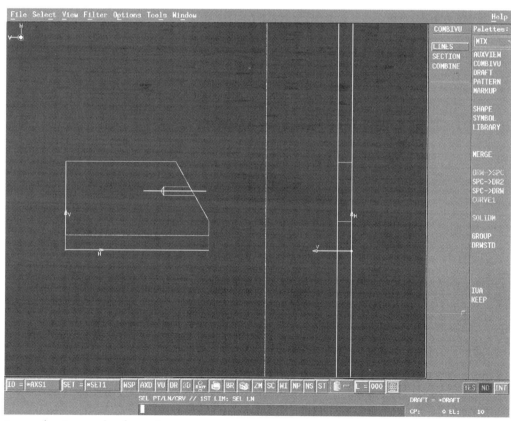

Screen showing results of step 4.

Chapter 6: Creating Multiple Views

5. At this point, you need to select the lines that will limit the next line to be created, the upper line forming the lower leg. Select the left line and the middle line vertical to the screen. You have now defined the limits. Select line 3 in the front view. Click on the YES button to end the limit selection.

6. To create the last line required for the angle section, select the outer lines vertical to the screen, and then select line 4 in the front view. Click on the YES button to end the limit selection.

7. Apart from the hole, the only remaining work is to relimit the corners using LIMIT1 > RELIMIT | TRIM ALL. (Of course, you could have created all the lines in the new view as unlimited lines, in which case continued selection of limiting lines would not be necessary.)

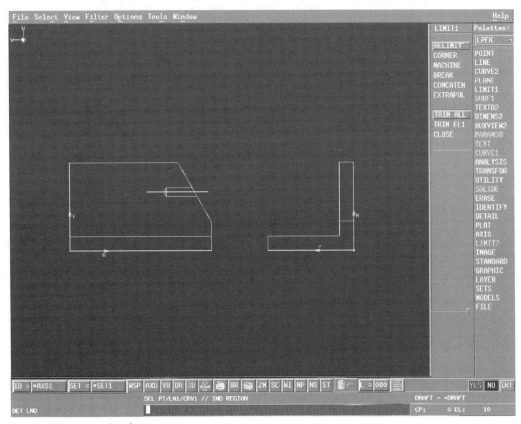

Screen showing results of step 7.

Exercise 1: Using AUXVIEW and COMBIVU

8. To create the hole, you must first define its center point. To create the first center line, select COMBIVU > LINES. Next, simply choose the center line of the hole in the front view, which will be projected across into the side view. To create the other center line you can use LINE > PARALLEL | UNLIM. Select in turn the lines on both sides of the hole in the side view.

9. In the message area, you are given the option of keying in the number of lines you require between the two selected lines. Since in this case you require only one line, key in *1* to the input information area. The center of the required hole lies on the intersection of the two lines. To create a point on this intersection, use POINT > PROJ INT | SINGLE | LIM. Create the three-diameter hole with CURVE2 > CIRCLE | DIAMETER.

NOTE: *In Chapter 1 you set the implicit points option in the standard settings to NO. If you now change this option to YES, the standard settings are obtained by clicking on the STD button in the fixed menu area. Having changed the implicit point setting, the point just created would be erased. To define the circle center with the YES option active, you merely position the cursor over the intersection of the lines and press mouse button 1. Once you have experimented with implicit points, reset the option to NO in the standard settings.*

10. The lines created for the circle center could be relimited, and then you could change them graphically with the GRAPHIC function. Instead, erase the two lines and recreate the center lines using the MARK UP function.

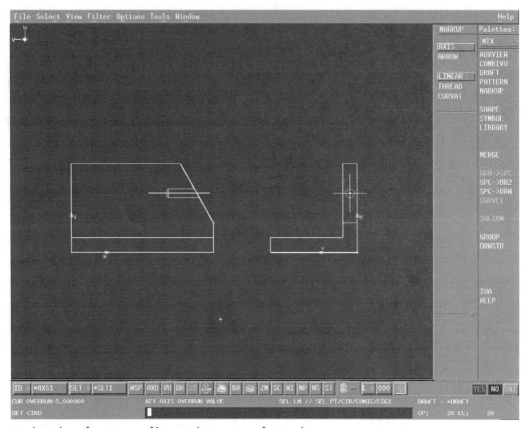

Final product after erasure of lines and recreation of center lines.

Finishing Details

Now that the side view is complete, it is time to return to the front view. You will change the lines of the hole to dotted line type, and the center line to dot dashed and thickness 1.

1. Select GRAPHIC > MODIFY | SAME. This option allows you to choose a reference element and then to change any element subsequently selected to the same graphical standards. Select one of the mark up center lines in the side view as your reference, and then select the hole center line in the front view. Note that the line type and thickness have changed.

Exercise 1: Using AUXVIEW and COMBIVU   107

2. To change the hole outlines to dotted, select GRAPHIC > MODIFY | CHOOSE | LINETYPE | DOTTED and select the five lines that comprise the hole.

3. Select GRAPHIC > MODIFY | CHOOSE | THKNESS | 1, and select the same five lines to change thickness from 4 to 1.

Your drawing should now resemble the following illustration, except for the line designations which will be used as references in the creation of the plan view.

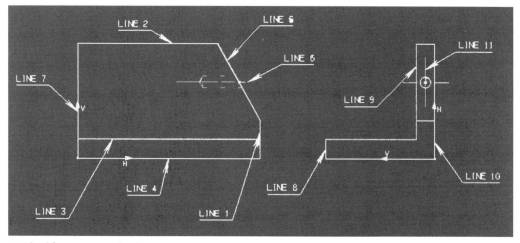

Finished front view and side view.

When working with the COMBIVU function you may see ELEMENTS IN CURRENT VIEW in the message area. This message means that the view you believed to be current is not. To determine the identity of the current view, and to change views as necessary, select AUXVIEW > CHANGE. After selecting this option, the current working view will be dimmed. The side view should be the current working view. If it is not the current view, select any element in the side view to make it current.

Another method of switching views is to click on the VU button in the fixed menu area. Again the current view will be shown dimmed and you can select any element in the other view to make it the current view.

Translating Views Around the Screen

Before moving on to create the plan view, a brief exercise in translating an individual view around the screen will be useful.

1. Select AUXVIEW > MODIFY | TRANSLAT. The side view should now be highlighted; if it is not highlighted, select the side view.

2. Using mouse button 2, indicate anywhere in the side view, and then indicate anywhere on the screen. The side view will move to the new position on the screen.

3. To move the view in a vertical direction only, select a line horizontal to the screen and indicate anywhere on the screen. The same procedure applies to moving the view in a horizontal direction only: select a line vertical to the screen and indicate anywhere on the screen.

4. Now that you have moved the view away from the original position, you need to move the side view back in line with the front view. Using AUXVIEW > MODIFY | TRANSLAT, select the V axis in the side view. Upon verifying that the side view is highlighted first, select the H axis in the front view. The two views will now be aligned vertically.

5. To move the side view in a horizontal direction, select the H axis in the side view and then indicate a new position for this line to the right of the front view. Your side view should now be back in its correct position.

When you have created a few more views, consider returning to AUXVIEW > MODIFY | TRANSLAT and experiment with moving the views around.

Plan View

Once the plan view is created you can create all the new geometry in this view without having to key in dimensions to the input information area. The steps to create the new view follow.

1. Select AUXVIEW > CREATE | NEW BGD. The first step is to define the plane on which the new view will be drawn. In this instance, it is line 4 in the front elevation view, so select line 4.

Exercise 1: Using AUXVIEW and COMBIVU

2. INDICATE POSITION YES:SAME ORIGIN appears in the message area. To ensure that the views use the same origin system and are properly related to each other, it is important to use the YES option here. Click on the YES button.

> **NOTE:** *Because the second view to be created will also use the third angle, defining the projection convention is not necessary. When you proceed to the next step, the default convention will be used.*

3. Indicate placement of the new view by indicating above the front view with mouse button 2. When asked to name the view, key in *PLAN*.

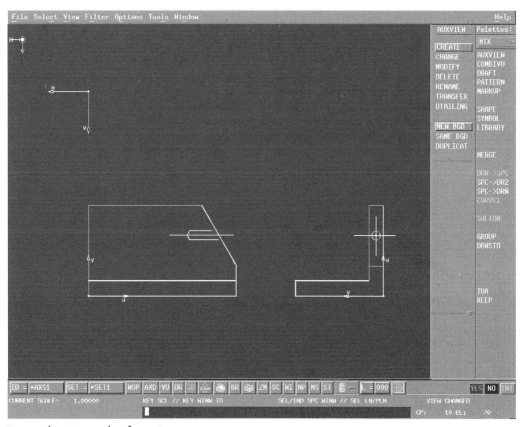

Screen showing results of step 3.

4. A new axis system appears. Note that the vertical axis has been inverted. Consequently, when you require a line to be horizontal to the screen in the new view you will need to remember that positive values would place the line below the H axis and negative values would place the line above the H axis.

Now that the new view has been created you need to create some geometry.

1. Because you can use the geometry from the other two views to create all the geometry in this new view, the LINE function is not required to create lines. Instead, select COMBIVU > LINES. To create all the geometry in the plan view, select lines 8, 9, and 10 in the side view, and lines 1 and 7 in the front view.

2. To create the short line in the plan view, select lines 2 and 5 in the front view to provide the intersection line. At this point, you have all lines in the plan view required to create the target drawing, except for the hole.

3. Relimit the lines using LIMIT1 > RELIMIT | TRIM ALL. Use LIMIT1 > RELIMIT | TRIM EL1 to relimit the intersection line.

4. To create the hole (elliptical rather than circular in the plan view), select COMBIVU > COMBINE. Defining the face on which the hole lies is required. Select line 5 in the front view.

5. At this juncture, you must select the hole from a view where it is shown as a true circle. Select the circle in the side view. The ellipse appears in the plane view.

6. Use MARK UP > AXIS to create the hole center lines.

✓ **TIP:** *If you cannot create any of the lines or the circle in the plan view, verify that the plan view is the current view using AUXVIEW > CHANGE or the VU button.*

Exercise 1: Using AUXVIEW and COMBIVU

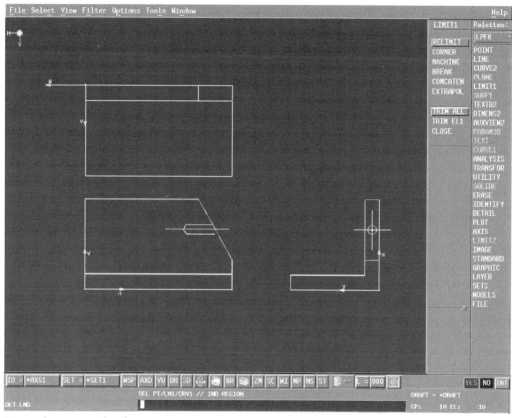

Screen showing results of step 3.

Creating holes that are not true in a view is obviously a very useful function. But what happens if you do not have a true view of the hole from which to create an ellipse? Consider creating a view containing a true view of a hole for the exclusive purpose of generating an ellipse in another view. You can always delete views when they are no longer useful.

Your drawing should now resemble the following illustration, excepting the line designations to be used in the creation of the isometric view.

112　　　　　　　　　**Chapter 6: Creating Multiple Views**

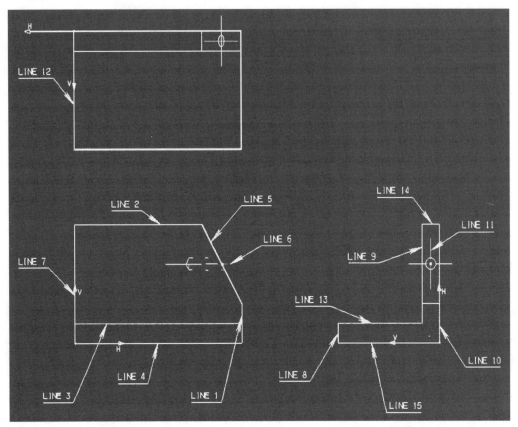

Finished front, side, and plan views.

Isometric View

This view can also be created without keying dimensions to the input information area.

1. Select AUXVIEW > CREATE | NEW BGD. Instead of selecting the defining plane, you need to key *XYZ* (the axis system for 3D work) into the input information area, which instructs CATIA that you are creating an isometric view.

2. When asked to define the position for the new view, use mouse button 2 to indicate a convenient position above the side view.

Exercise 1: Using AUXVIEW and COMBIVU

3. You will now be asked to name the view. Give the view a meaningful name, such as *ISO*. At this point, you will see the XYZ axis as well as the V and H axes.

4. To create geometry in the view, use COMBIVU > COMBINE rather than COMBIVU > LINES. The COMBINE option allows you to combine the information available in the other views to create the isometric view. When asked to define a plane or curve on which the geometry of other views lies, select line 10 in the side view. Next, to select geometry from the other two views which lie on this plane, select lines 1, 2, and 5 from the front view. Three lines now appear in the isometric view. Because these three lines are the only lines lying on the selected plane, click on the YES button to end the selection.

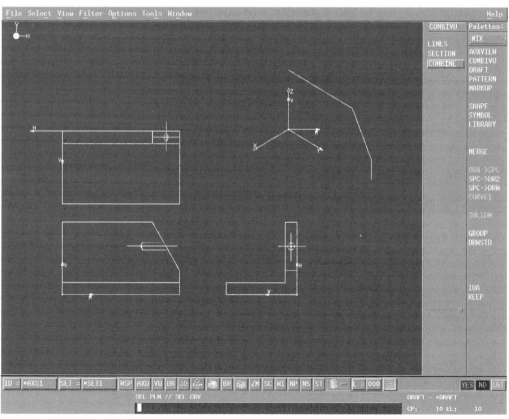

Screen showing results of step 4.

Chapter 6: Creating Multiple Views

5. You now need to define another plane on which to create more geometry. Select line 12 from the plan view. Select lines from the side view to create lines on the selected plane: choose lines 8, 9, 13, and 14. Another four lines appear in the isometric view. Click the YES button to end the selection.

6. The next plane to use to create geometry could be line 9 in the side view. After selecting line 9, select lines 1, 2, 3, and 5 from the front view. Click on the YES button.

7. One of the corners must now be relimited. Select LIMIT1 > RELIMIT | TRIM ALL, and select the two lines created from lines 1 and 3.

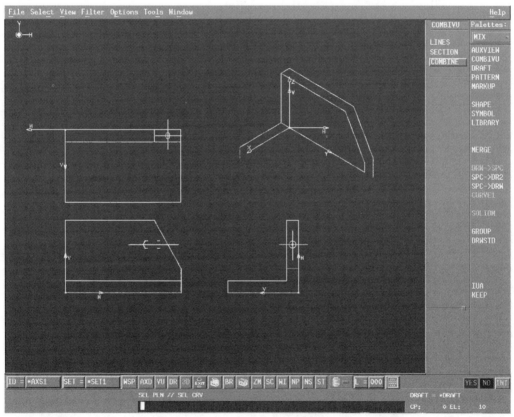

Screen results of step 7.

Exercise 1: Using AUXVIEW and COMBIVU

8. To continue creating the isometric view select COMBIVU > COMBINE. The next plane to be used could be line 1 from the front view. After selecting line 1, select lines 8, 13, and 15 from the side view. The rest of the lines required for the isometric view can now be created using LINE > PT-PT | SEGMENT.

9. At this point, the only remaining feature to be created is the circle. Select COMBIVU > COMBINE. To define the plane on which the hole lies, select line 5 from the front view. To define the hole you need to select the circle from a view in which it is shown true, select the circle in the side view. The ellipse appears in the isometric view. The isometric view is complete.

Your drawing should now resemble the target drawing. If the chamfer on the angle section had been a radius corner as shown in the following illustration, you could have created the isometric view in the same manner. But you would have had to select the curved face rather than the angled face when you indicated the face on which the hole was defined.

Equal angle section with radius corner.

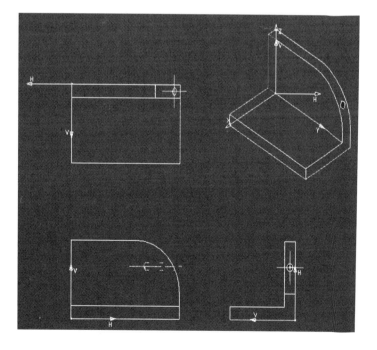

Exercise 2: Using AXIS in DRAW Mode

The AXIS function has many uses. This exercise explores AXIS used with TRANSFOR > MOVE to create combined translations and rotations, and with the ANALYSIS function to provide additional analysis results.

Each view in Exercise 1 was created with its respective axis. In the following steps, an additional axis is created. The new axis is used to perform a translation and rotation in a single step.

1. Read the model created in the angle section of Exercise 2 in Chapter 3 with FILE > READ. To create a second axis in the model (without an additional view), produce a point to locate the axis. Select POINT > COORD | SINGLE, and create the point at coordinates X = 60 and Y = 30.

2. Select AXIS > CREATE. The first message asks you to select a point to define the position of the new axis. Select the point just created.

3. The next message asks you to define the angle of the new axis by selecting a line at the required angle or entering the angle value. In this exercise the angle is provided; key in *30* and press <Enter>. The new axis will be on the point and with the horizontal axis at an angle of 30 to the horizontal axis of the original axis. This new axis will be the current axis system.

Exercise 2: Using AXIS in DRAW Mode

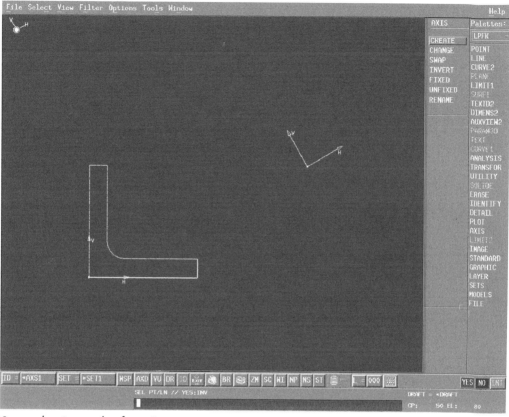

Screen showing results of step 3.

The next step in this exercise is to duplicate the angle section on the new axis system. With the use of the TRANSFOR function, the translation could be achieved by creating a translation and duplicating the angle, and then rotating the new angle through 30 degrees. A faster way to achieve the same result is to use the TRANSFOR function's MOVE option described in Chapter 5 to create a transformation between two axis systems.

1. Select TRANSFOR > CREATE | MOVE. When asked to select the first axis, select the original axis system.

2. When asked to select the second axis, select the new axis system. The displayed vectors include a translation and a rotation.

Chapter 6: Creating Multiple Views

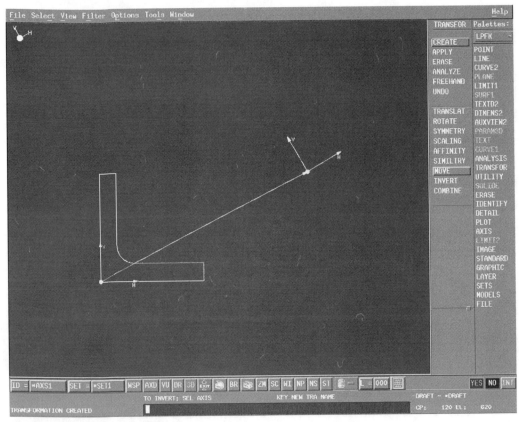

Screen showing results of step 2.

3. Select the APPLY option from the TRANSFOR function menu without leaving the function. Select DUPLICATE from the APPLY window, and then select the angle elements. After the elements have been selected, you will have achieved the target drawing.

4. Save the model.

Exercise 2: Using AXIS in DRAW Mode

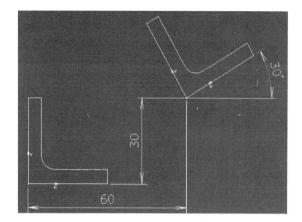

Duplicated angle section.

Multi-selection Options

Selecting all elements one by one can be a time-consuming process, especially if your drawing contains numerous elements. An alternative method of selecting all the elements is to use multi-selection options. Three such options that could have been used in the previous exercise are listed below.

- Select all elements in DRAW by using *DRW.

- Select all elements by trapping them with *TRP.

- Select all line elements by using *LND, and then select the curve by using *CIRD. These multi-selection options could be used at the same time by entering *LND+*CIRD.

To experiment with multi-selection options, use the model in the above exercise, and erase the second angle section.

- After recreating the MOVE transformation and selecting the APPLY option with DUPLICATE, key in *DRW to the input information area and press <Enter>. All elements in the current view will be copied as defined by the MOVE transformation.

- Follow the procedure in the first bulleted item, but use *TRP instead of *DRW. When asked to indicate the first point to define a box in which to trap all the elements, choose a point at the left of the top left of the angle section. The second point could be to the right of the bottom right of the angle section. You will then be given the option of closing the trap. Click on YES. All elements within the trap are highlighted. Click on YES again to confirm the trap. All elements are copied to their new positions.

- Follow the procedure in the first bulleted item, but use *LND* instead of *DRW*. Note that only the line elements have moved. Enter *CIRD*; the curve has moved.

- Follow the procedure in the first bulleted item, but use *LND+*CIRD* instead of *DRW*. Both the line elements and the curve have moved.

Exercise 3: Using AXIS with ANALYSIS

This exercise is focused on the effect of a two-axis system when using ANALYSIS.

1. Read the model created in Exercise 2 (duplicated angle section) with FILE > READ. Before using the ANALYSIS function, you need to identify the current axis system.

2. Select AXIS > CHANGE. When this function and option are selected all draw elements are dimmed, and are nonselectable. The only selectable objects are the two-axis systems. The current axis system displays with solid lines, and the other with dashed lines. If the original axis displays with dashed lines, select the axis to change it to the current system. If the original axis is already solid, proceed to step 3. Checking and changing of axis systems can also be carried out by using the AXD button on the fixed menu.

3. Select ANALYSIS > NUMERIC. The position of any element subjected to analysis is related to the current axis.

4. Make the new axis the current one. ANALYSIS results will now be relative to the new axis system.

Exercise 3: Using AXIS with ANALYSIS

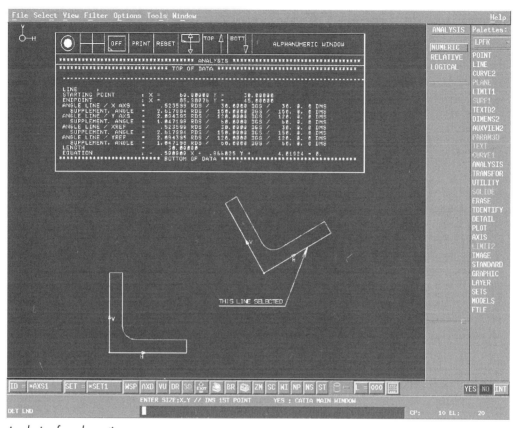

Analysis of angle section.

At this point, you could create shapes from the two-angle sections. (Refer to Chapter 4 for direction if necessary.) Analyze the shapes in relation to the two different axis systems.

There are two major benefits derived from the ability to define the axis about which analysis is performed. First, assume that you require analysis of the center of gravity (or any other feature) relative to a particular point. Extra calculations are not necessary because you would only need to create an additional axis system and perform the analysis on that axis.

Another benefit of using a two-axis system is that if the new axis system is current, as in the above exercise, any new lines parallel to the sides of the angle could be created using horizontal and vertical lines. This would be useful if you are working with a large amount of geometry at an odd angle to the original axis system.

Exercise 4: Using DRAFT and AUXVIEW

The DRAFT function is introduced here with a simple exercise. A more detailed exploration of DRAFT usage appears in Chapter 7, "Annotating Drawings." With the use of the DRAFT function, a model can have multiple views, but the user has the opportunity to visualize and plot only selected views.

1. Read the model created in Exercise 1 using FILE > READ (i.e., the equal angle section views and the isometric view). Select DRAFT > CREATE. When asked for the name of the new draft in the message area, key in *NEW DRAFT* to the input information area and press <Enter>. The new draft is created.

2. To demonstrate how DRAFT works, you will delete one of the views in the model. Select AUXVIEW > DELETE. Select the isometric view as the view to be deleted. Click on the YES button to confirm view deletion. The view is deleted from the draft called *NEW DRAFT.*

3. Select DRAFT > CHANGE. Click the YES button to list all available drafts available, even though in this case you only have one alternative draft. Select **DRAFT* from the resulting list. The isometric view has returned. (The isometric view was only deleted from the draft called *NEW DRAFT.*)

Exercise 4: Using DRAFT and AUXVIEW   123

DRAFT (Default DRAFT).

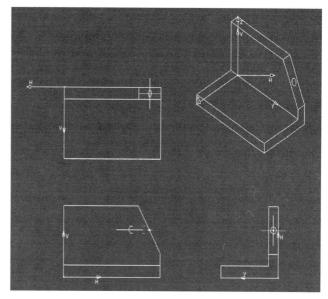

4. You could now swap between the two drafts using DRAFT > CHANGE to view the differences. You could also delete additional views from each draft and examine the effects on different drafts. This facility could therefore be used to create different drawings from the same model using common views, but also using views specific to different drawings.

5. Save the model.

NEW DRAFT (draft created in exercise).

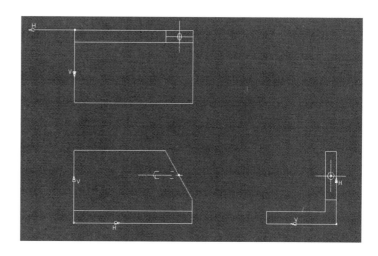

Summary

Topics covered in this chapter included creating multiple views in a model, using elements in one view to create elements in another, creating multiple drafts in a model, and creating multiple axis systems in a model.

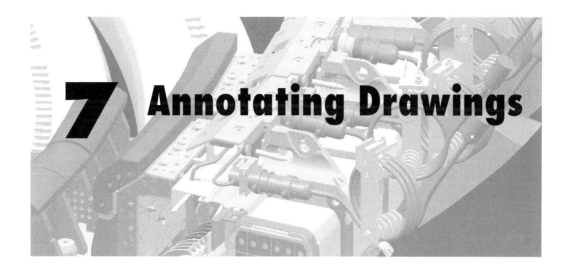

7 Annotating Drawings

This chapter is focused on the creation of finished engineering drawings for manufacturing. The following four functions, all accessed via palette menus, will be introduced:

- TEXTD2. Adding text to a drawing
- DIMENS2. Adding dimensions to a drawing
- PATTERN. Placing cross hatching on a section view
- DRWSTD. Managing text, dimension, and pattern descriptions as well as company standards

The DRWSTD function is typically the purview of system administrators. Chapter topics also encompass additional details on the usage of DETAIL, LIBRARY, SYMBOL, and DRAFT.

TEXTD2

The TEXTD2 function is available only in DRAW mode, although a TEXT function is available in SPACE mode and is covered in Chapter 16. The TEXTD2 function is used to create the items described in the next table.

TEXTD2 CREATE Options

Option	Description
CREATE | TEXT | SIMPLE	Simple text.
CREATE | TEXT | LEADER	Text with a leader arrow.
CREATE | TEXT | FITTED	Text fitted between two points.
CREATE | TEXT | FIT_LEAD	Text fitted between two points and with a leader.
CREATE | TEXT | ROUGHNESS	Roughness (machining) symbols.
CREATE | GEOM TOL	Geometrical tolerance symbols.
CREATE | DAT FEAT	Datum features.
CREATE | DAT TARG	Datum targets.
CREATE | BALLOON	Balloons.

The following illustrations present samples of text and symbols that can be created with the TEXTD2 function.

Examples of text and symbols created using TEXTD2.

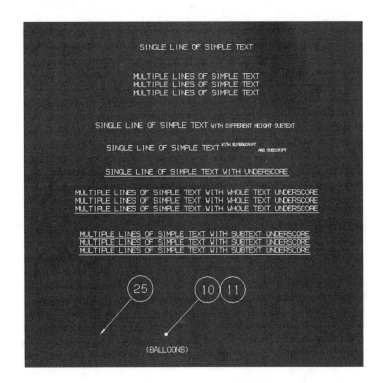

More examples of text and symbols created using TEXTD2.

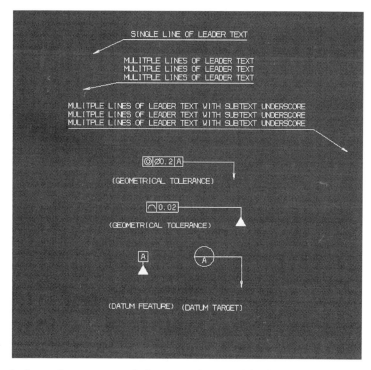

Once text or symbols have been created they can be modified in many ways. MODIFY options are listed in the next table.

TEXTD2 MODIFY Options

Option	Description
MODIFY	TEXT	EDIT	Edit text.
MODIFY	TEXT	GRAPHIC	Change graphical representation of text.
MODIFY	LEADER	Add, delete, or change path(s) of leader.
MODIFY	SYMBOL	Change symbol.
MODIFY	ORIENTAT	Change text orientation.
MODIFY	LOCATION	Change location of text or symbol.
DELETE	Delete lines of text.
MANAGE	Modify text visualization.

128   *Chapter 7: Annotating Drawings*

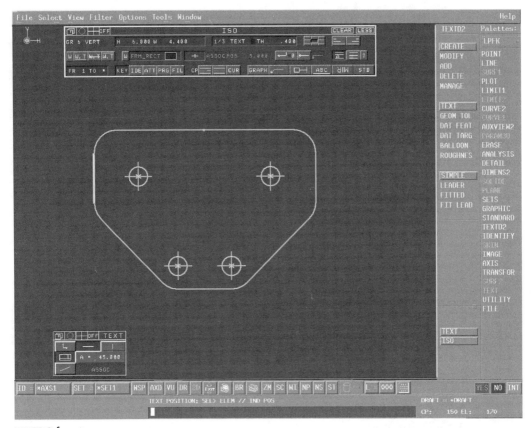

TEXTD2 function menu.

When the TEXTD2 function is in use, a Management dialog appears at the top of the screen. To expand the window select the MORE icon at top right; to return the window to its original size select the LESS icon. The window is used to define the presentation of text on the screen. The various buttons and respective functions in TEXTD2 are described below.

TEXTD2

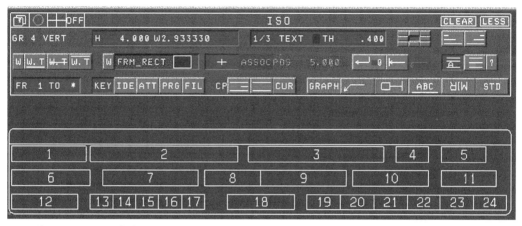

Typical TEXTD2 expanded Management dialog.

➥ **NOTE:** *The upper half of the previous illustration shows the TEXTD2 expanded Management dialog. The lower half of the illustration is a key to the buttons described in the following list. The window on your screen may not match the illustration. The DRWSTD function is used to set the configuration for the window; access to window modification commands may be limited to your system administrator.*

- Button 1—Graphism name. Defines the font style used. Upon selecting the letters in white, a list displays available text fonts. You determine the current font by making a selection in the list.

- Button 2—Height and Width. Defines the height and width of selected font. The width option may or may not be selectable depending on the font defined in the DRWSTD function. This option should be set to 4mm for the exercises in this chapter.

- Button 3—Thickness and Color. Defines the thickness and color for the text, score, frame, leader, and symbol used. The TH box refers to the thickness of the various elements of text. In this chapter, the text will be 0.4mm thick. If you select the second number, the first number and the word near the numbers change to allow you to modify the thickness of various text elements. If you select the word next to the numbers (i.e., TEXT), a window for defining text color appears.

- Button 4—Anchor point. Defines anchor point for text relative to the selected insertion point. This option should be set to the central button of the set of nine for subsequent exercises.

- Button 5—Justification. Defines text justification. Selecting the button at the left will produce left justified text, and the button at the right, right justified text. If neither button is selected, the text will be center justified. Choose the latter option for the following exercises.

- Button 6—Score. Defines position of scoring. Scoring can be under, through, or over the text. You also have a choice of whether scoring is applied to all text entered or to an individual piece of subtext. The W button refers to "whole" text and the S button refers to "subtext." The W/S button is toggled. Verify that scoring options are not selected for purposes of the exercises.

- Button 7—Framing. Defines framing in DRWSTD, such as box, circle, and arrow, among others. Once again you have the option of applying a frame to whole or subtext. To change the frame in use, select the frame symbol and a list of available frames displays. Verify that no frames are selected prior to commencing the exercises.

- Button 8—Associative text. Links text to a particular element. This option should not be selected for the following exercises.

- Button 9—Associative text positioning. Defines the distance of text from an associated element. You can change this dimension only if the Associative text button is selected.

- Button 10—Edit. Allows you to determine whether a new line of text starts after, below, or at the beginning of existing text. None of these options should be selected before beginning the exercises.

- Button 11—Text grid. Defines creation of subscript and superscript text.

- Button 12—Format. Defines length of entered text.

- Button 13—Key text. Permits text to be keyed in or copied from existing text. This option should be selected prior to the exercises.

- Button 14—Identification. Allows CATIA element identification to be used for text purposes.

- Button 15—Attribute. Allows attributes of a selected element to be used for text purposes.

- Button 16—Program. Allows text to be created from the results of an external calculation program.

- Button 17—File. Allows text to be taken from an external file.

- Button 18—Copy mode. Allows the user to copy existing text.Button 19—Graphic parameter. When selected, accesses window through which user can alter the graphical display of the text.

- Button 20—Leader parameter. When selected, accesses window through which user user can set text-leader relationships.

- Button 21—Datum parameter. Selection accesses a window through which user can change datum symbol.

- Button 22—Score parameter. Selection accesses a window through which user can set text-score relationships.

- Button 23—Mirroring parameter. Selection accesses a window through which user can create mirrored text.

- Button 24—Standard parameter. Selection accesses a window through which user can alter standard text.

The next illustration shows one of the additional parameter windows for button 22. The window you see may differ from the illustration, depending on how the DRWSTD function is set by your system administrator.

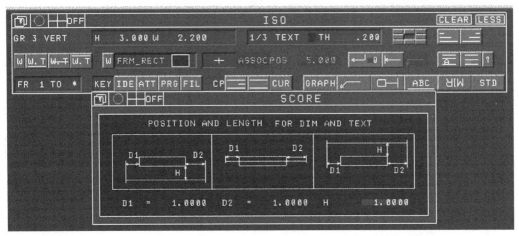

Typical text parameter window.

When using TEXTD2, note that a small window at the bottom left of the screen appears in addition to the Management dialog at the top. The Orientation Options window is used to define text orientation. The following list describes the text orientation options.

TEXTD2 Orientation Options window.

- Button 1—Positions text relative to the current view.

- Button 2—Positions text relative to the screen.

- Button 3—Positions text relative to a selected element.

- Button 4—Positions text on the horizontal relative to the current view or screen, depending on whether button 1 or 2 is selected.

- Button 5—Positions text on the vertical relative to the current view or screen, depending on whether button 1 or 2 is selected.

- Button 6—Number entered here defines the angle of the created text if button 3 is selected; if zero is entered the text will be parallel to the selected element.

- Button 7—Text is associated with a reference element.

DIMENS2

The DIMENS2 function is available only in DRAW mode. CREATE suboptions are described in the next table.

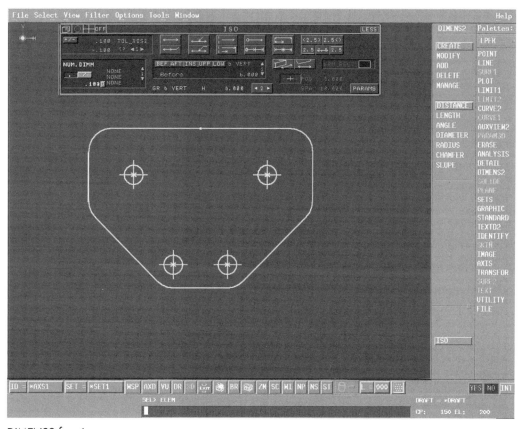

DIMENS2 function menu.

DIMENS2 CREATE Options

Option	Description
CREATE	DISTANCE	Dimensions between elements.
CREATE	LENGTH	Dimensions on the length of an element.
CREATE	ANGLE	Angular dimensions between elements.
CREATE	DIAMETER	Diameter dimensions on curves.
CREATE	RADIUS	Radius dimensions on curves.
CREATE	CHAMFER	Dimensions on chamfers.
CREATE	SLOPE	Dimensions on slopes.

Additional options for the seven suboptions described above are also available. The following list describes options for creating distance dimensions. Similar options are available for the remaining six options in the table.

- Simple dimensions between two elements; dimension is on the line.

- Simple dimension between two elements; dimension is connected to the dimension line with a leader. This leader can be from the center of the dimension line or the end.

- Dimensions (in line or staggered) can be taken from a datum point.

- Cumulative dimensions can be created.

- Only one end of a dimension line is shown (half dimension).

The next illustrations demonstrate some of the dimensions that can be created with the DIMENS2 function.

DIMENS2

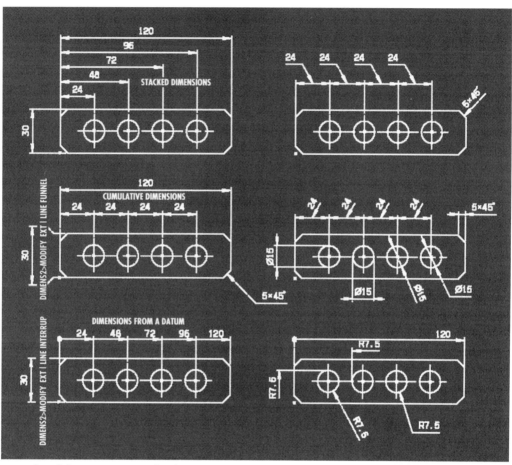

Examples of dimensions created with DIMENS2.

136 Chapter 7: Annotating Drawings

Examples of dimensions and tolerances created with DIMENS2.

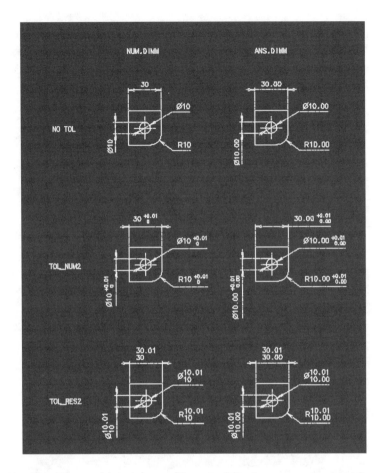

Once dimensions have been created they can be modified in many ways. The next table describes MODIFY options.

DIMENS2 MODIFY Options

Option	Description
MODIFY | DIM LINE	Move dimension line.
MODIFY | VALUE | LOCATION	Move dimension value.
MODIFY | VALUE | FAKE DIM	Change dimension value.
MODIFY | VALUE | TRUE DIM	Restore fake dimension value.
MODIFY | VALUE | NUM DISP	Modify character graphic dimension value.

DIMENS2

Option	Description
MODIFY | VALUE | GRAPHIC	Modify graphical representation of dimension value.
MODIFY | EXT LINE | LENGTH	Change length of dimension extension line.
MODIFY | EXT LINE | SLANT	Change angle of dimension extension line.
MODIFY | EXT LINE | ERASE	Delete dimension extension line.
MODIFY | EXT LINE | INTERRUP	Break dimension extension line.
MODIFY | EXT LINE | FUNNEL	Modify dimension extension line to create funnel type extension line.
MODIFY | EXT LINE | RESTORE	Restore modified dimension extension line to original state.
MODIFY | SYMBOL | CHANGE	Change dimension line end symbol.
MODIFY | SYMBOL | INVERT	Invert dimension line end symbol.
MODIFY | TOL | CHANGE	Add or modify tolerance.
MODIFY | TOL | DELETE	Delete tolerance.
MODIFY | ADD	Add dimension.
DELETE	Delete dimension without changing to ERASE function.
MANAGE	Modify visualization of dimensions.

Upon using the DIMENS2 function, a Management dialog appears at the top of the screen. To expand the window select the MORE icon at the top right; to restore the window's original size select the LESS icon. This dialog is used to define the presentation of dimensions on the screen. The buttons are described below.

Chapter 7: Annotating Drawings

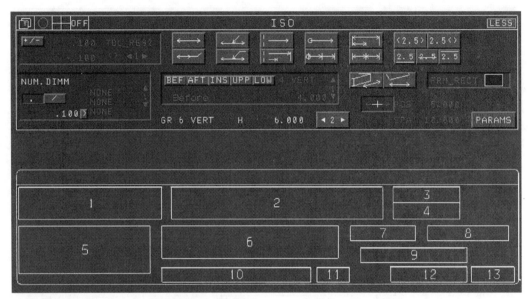

Typical DIMENS2 expanded Management dialog.

> **NOTE:** *The DIMENS2 expanded Management dialog on your system may not precisely match the above illustration. The window configuration is set via the DRWSTD function; configuration modification may be available only to your system administrator.*

- Button 1—Tolerancing mode. Switches tolerancing mode on or off, and defines tolerances. This option should be switched off prior to beginning the exercises.

- Button 2—Presentation mode. Defines presentation of dimensions (i.e., single, cumulative, from a datum or stacked).

- Button 3—Dimension text location mode.

- Button 4—Score line. Defines position of score. Score can be under, through, or over text.

- Button 5—Numerical display. Defines dimension text display standard to be set; also used to define the accuracy of the displayed dimension.

- Button 6—Associated text. Allows use and position of associated text to be defined.

- Button 7—Calculation mode.

- Button 8—Framing. Defines framing as set in DRWSTD (i.e., box, circle, and arrow, among others). To use a frame, select the frame shown and the frame description will become selectable (i.e., framing is switched on). To change the frame in use, select the frame description you want and a list of available frames will display. No frames should be selected prior to the exercises.

- Button 9—Dimension position mode.

- Button 10—Graphism. Defines font style and text height. Set to 6mm for the following exercises.

- Button 11—Dual mode. Defines use of dual dimensioning.

- Button 12—Spacing mode. Defines the distance for automatic spacing of dimension lines when using the stacking presentation option (see Button 2).

- Button 13—Access to parameters. Upon selecting, user can modify dimensioning parameters.

The following illustration displays available parameters. Your options may differ because window contents are typically controlled by a system administrator via the DRWSTD function. The illustration demonstrates the vast number of standards and permutations thereof in DIMENS2. The option shown is DIM TEXT.

Chapter 7: Annotating Drawings

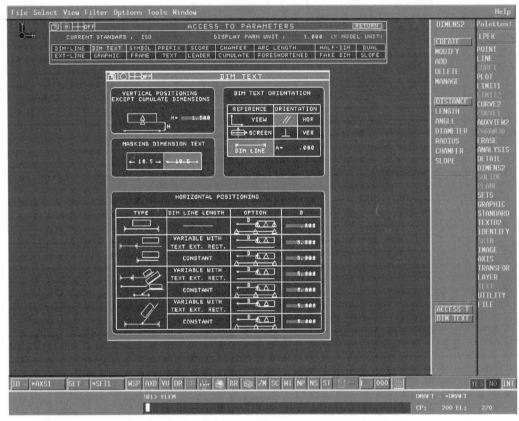

Typical DIMENS2 Parameter Management dialog.

Descriptions of parameter options that may be available to you appear in the next table.

Option	Description
DIM-LINE	Defines orientation, positioning, and length of dimension lines.
EXT-LINE	Defines orientation, positioning, and length of extension line.
DIM TEXT	Defines orientation and positioning of dimension text relative to the dimension lines. (This parameter window is shown in the previous illustration.)
GRAPHIC	Defines display, thickness, and color of all dimension elements.
SYMBOL	Defines symbols used in dimensioning.

Option	Description
FRAME	Defines size of framing.
PREFIX	Defines prefix used in various dimensions (e.g., diameter, radius, slope).
TEXT	Defines positioning of text relative to dimension lines.
SCORE	Defines size and position of scoring.
LEADER	Defines leader end symbol, leader length, and leader orientation.
CHAMFER	Defines how a chamfer dimension is displayed.
CUMULATE	Defines display and orientation of cumulative dimensions.
ARC LENGTH	Defines display of arc dimensions.
FORESHORTENED	Defines display of foreshortened radius dimensions.
HALF-DIM	Defines display of half-dimension mode.
FAKE DIM	Defines display of fake dimensions.
DUAL	Defines display of dual dimensions.
SLOPE	Defines display of slope dimensions.

NOTE: As mentioned previously, certain parameter options and suboptions may not be available because they are controlled by system administrators.

PATTERN

The PATTERN function is available only in DRAW mode and is used to create and arrange patterns in diverse ways. Selected options are described in the next table.

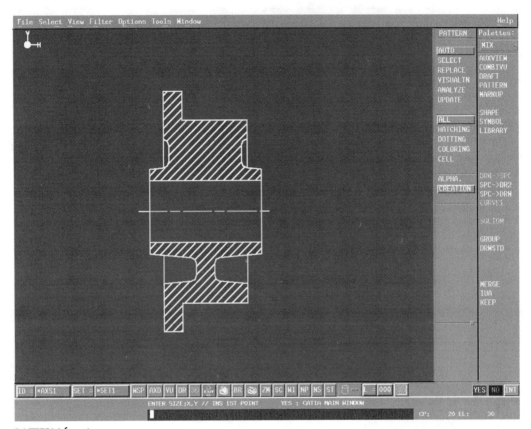

PATTERN function menu.

Option	Description
AUTO or SELECT	HATCHING	Create hatching pattern to fill closed areas.
AUTO or SELECT	DOTTING	Create dot pattern to fill closed areas.
AUTO or SELECT	COLORING	Create colored pattern to fill closed areas.
REPLACE	Replace with different pattern.
VISUALTN	Change visualization.
ANALYZE	Analyze to identify pattern.
UPDATE	Update pattern when pattern boundaries change.

As discussed in Chapter 4, patterns can also be analyzed using the ANALYSIS function. When employing ANALYSIS, you can obtain the pattern's area, perimeter, and center of gravity in the same manner a shape is analyzed in Chapter 4. In fact, pattern identity is the same as for a shape (i.e., *SHAP*).

DRWSTD (Draw Standard)

The DRWSTD function is used to manage and create descriptions and standards used in the TEXTD2 and DIMENS2 functions. It is also used to create patterns in the PATTERN function.

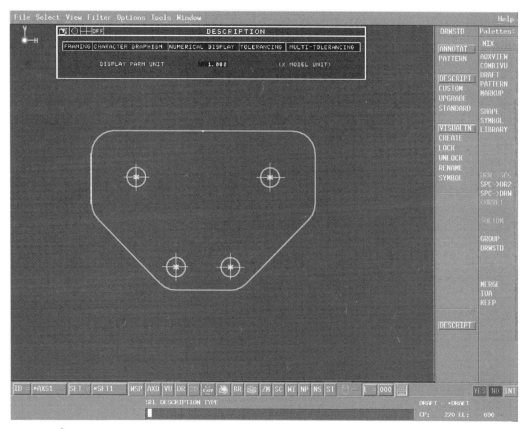

DRWSTD function menu.

Typically, the DRWSTD function is controlled by system administrators for purposes of creating and managing company standards. Because this function is not generally available to the average user, usage and option modification are not covered in this chapter. Instead, the function's purpose is summarized below.

- Visualize descriptions used in TEXTD2 and DIMENS2 functions.

- Create new descriptions.

- Create new patterns.

- Lock and unlock existing descriptions and patterns.

- Rename existing descriptions and patterns.

- Assign values to different types of end symbols used in TEXTD2 and DIMENS2 functions.

- Define the Management Dialog windows used in TEXTD2 and DIMENS2 functions.

- Create company standards for use in TEXTD2 and DIMENS2 functions.

- Define the procedure for converting and modification of old dimensions and text annotations.

- Define hatching, dotting, and coloring used in PATTERN function.

One option that should be available to average users in DRWSTD is the ability to view the available text characters, dimensioning standards, and tolerancing standards. Select DRWSTD > ANNOTAT | DESCRIPT | VISUALTN. The dialog box shown in the next illustration appears.

Description dialog.

DRWSTD (Draw Standard)

Upon selecting the NUMERICAL DISPLAY option in the DESCRIPTION window, a LIST window appears. Available standards for use in dimensioning are seen here, and likely vary from site to site.

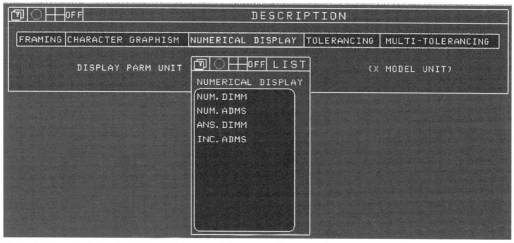

Typical Numerical Display List window.

The DRWSTD function is also used to define symbols at line ends for text and dimensions. To view the symbols, select DRWSTD > ANNOTAT | DESCRIPT | SYMBOL. After selecting this option sequence, a Symbol dialog appears. Symbol sizes are changed by entering new values in the column headed SIZE PARAMETERS. The current standard in use displays at the top of the dialog box (e.g., ISO). The symbol standard cannot be changed from this window.

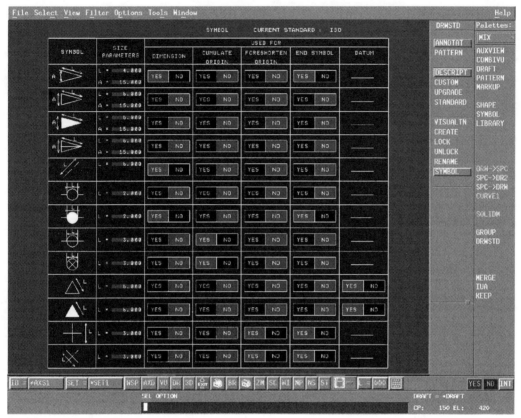

Typical Symbol dialog.

Exercise 1: Using TEXTD2 and DIMENS2

Creating a Drawing Sheet Blank

1. Read the model created in Chapter 3, Exercise 6 (*Side Plate*).

2. Create the second view by selecting AUXVIEW > CREATE | NEW BGD, and then select one of the vertical lines to define the new view. Click on the YES button to ensure that the new view is correctly associated with the first view. Because the views created in the Chapter 3 exercise were third angle projection, AUXVIEW will default to the last used choice of

Exercise 1: Using TEXTD2 and DIMENS2   147

view definition. You do not need to define whether you require first or third angle projection.

3. Indicate the position of the new view to the right of the front view. Name the view *SIDE*.

4. Using LINE > HORIZONT | UNLIM, create the two lines 10 apart. With COMBIVU project the top and bottom lines across into the new view.

5. With LIMIT1 > RELIMIT | TRIM ALL, trim the corners to provide finished geometry for the side view.

At this juncture, a drawing sheet blank will be created as a detail. Creating a blank as a detail enables you to use the detail more than once.

1. Select DETAIL > CREATE. Key in *A1 BLANK* for the detail name and press <Enter>.

2. Click on the YES button, or key in a comment if required. Because your workspace has changed, W/space : A1 BLANK should appear at screen bottom right.

→ **NOTE:** *If you are not in the A1 BLANK workspace select DETAIL > CHANGE and press <Enter>.*

Overall dimensions of A1 blank.

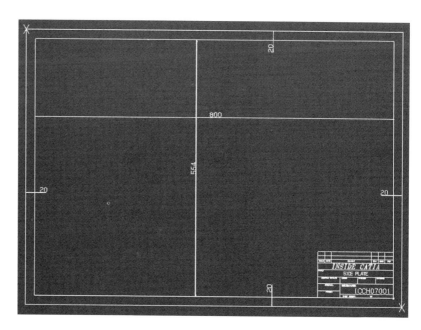

The first steps involved in creating the drawing sheet follow.

1. Create the inner boundary to the dimensions shown in the previous illustration. Select LINE > PT-PT | SEGMENT | HOR-VERT. You now need to key in the coordinates for the start and finish points of the boundary. Because the start point is on the axis, key in , and press <Enter>. For the finish point, key in *-800,554* and press <Enter>. You have created the lower horizontal line and the left vertical line.

2. To create the other two lines, select LINE > PT-PT | SEGMENT | VERT-HOR. Key in the same coordinates as in step 1 for the start and finish points of the boundary lines: , and *-800,554*. The inner boundary is complete with the insertion of the right vertical line and the top horizontal line.

3. To create the outer boundary, select LINE > PARALLEL | UNLIM. Select a line and indicate the side on which you want the parallel line; key in *20* and press <Enter>.

4. Repeat step 3 with the other three lines. Use LIMIT1 > RELIM | TRIM ALL to relimit the corners. The outer boundary could also have been created using LINE > PT-PT | SEGMENT | VERT-HOR or HOR-VERT in the same way as the inner boundary.

5. With POINT > PROJ/INT create a point in the top left corner and another in the bottom right corner. These points will be used when setting up plots in Chapter 8.

Creating a Title Block

The next stage in creating the drawing blank is to create the title block. The lines will be created first to the dimensions in the next illustration.

Exercise 1: Using TEXTD2 and DIMENS2

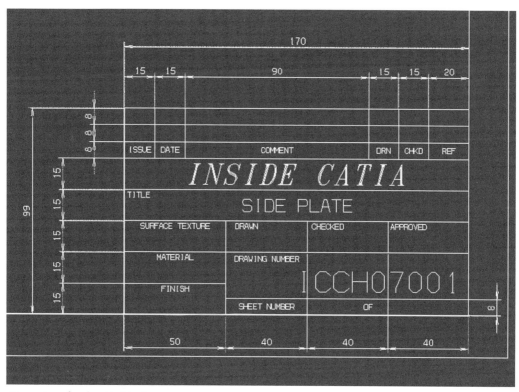

Dimensions for the title block.

1. Create the two boundary lines, 99 and 170 long, by using LINE > PARALLEL | UNLIM.

2. The inner boundary lines are used as references to create the parallel lines. Use LIMIT1 > RELIMIT | TRIM ALL to relimit the lines for the top left corner. With LIMIT1 > RELIMIT | TRIM EL1, relimit the two ends limited by the main boundary.

3. Create the three lines spaced at 8mm from the top line of the title block. To create these lines use LINE > PARALLEL | SEGMENT | ONE LIM. Select the top line, 170 long, and indicate below the line.

4. KEY DIST(,NUM) in the message area means that you can enter the distance from the first line as well as key in the number of required new lines that you require. (If only one line is required, you need not key in a number.) Key in *8,3* and press <Enter>. Three vector arrows appear.

5. At this point, you must define the length of the new lines. You have two choices: key in *170* or select the top 170 line. Create the three lines at 170 long spaced at 15 from the last created line using the same procedure seen in earlier steps.

6. To create the final few lines, you can use LINE > PARALLEL | SEGMENT | TWO LIM to avoid relimiting the lines after creation. To create the two vertical lines 40 apart, select the inner boundary vertical line and indicate to the left of the line. Key in *40,2* (2 lines 40 apart), and press <Enter>.

7. To define the two limiting lines, select in turn the two horizontal lines 15 apart that were created previously. The two created vertical lines are limited by two horizontal lines, 15 apart.

8. The remaining lines for the title block can now be created using LINE > PARALLEL | SEGMENT | ONE LIM or LINE > PARALLEL | SEGMENT | TWO LIM. An alternative method for creating all lines for the title block is using LINE > PARALLEL | UNLIM followed by LIMIT1 > RELIMIT to trim all the lines.

Annotating a Title Block

The last stage in creating the drawing blank is to annotate the title block. The text will be created and inserted as shown in the next illustration. After creating the text, all points will be placed in NO SHOW.

Exercise 1: Using TEXTD2 and DIMENS2

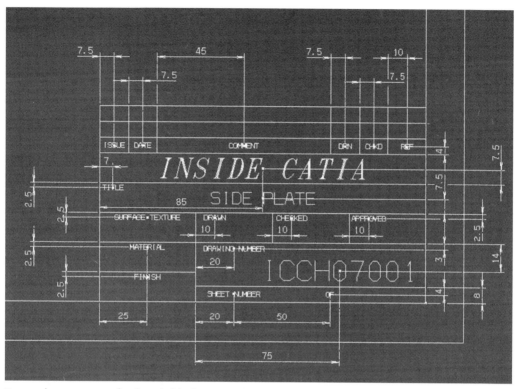

Text and text positions for the title block.

1. The first stage is to define the points required for insertion of the text. With POINT > INT/PROJ select the two lines that form the lower left corner of the ISSUE box. After creating this point, change the point function to POINT > COORD and select the point just created.

2. Key in 7.5,4 and press <Enter> to create the point required for insertion of ISSUE. (This point could have been created in other ways. For instance, with POINT > INT/PROJ you could have created points in diagonally opposite corners of the ISSUE box and then used the two points to create the point between them.)

3. Create the rest of the required insertion points in the same way the first was created.

4. To create the text for the title block, select TEXTD2 > CREATE | TEXT | SIMPLE. Set the parameters listed below.

- Button 1—Select the text font to be used for most of the text. A list of available fonts is accessed by selecting the font choice already in the box; you can then select an alternative from the list.
- Button 2—Input *2.5* to the height box.
- Button 3—Retain default.
- Button 4—Set to the middle box of the group of nine.
- Button 5—Neither of the justification boxes should be selected. The text will therefore be centered.
- Button 6—Do not select any of the score boxes. Check both the W and S boxes.
- Button 7—No framing should be selected.
- Button 8—No associative text should be selected.
- Button 9—The dimension here should not be selectable if associative text is switched off.
- Button 10—Neither of these two boxes should be selected at this stage.
- Button 11—Neither of these two boxes should be selected at this stage.
- Button 12—See the next illustration.
- Button 13—This button should be selected to permit entry of required text.
- Button 14—Deselect.
- Button 15—Deselect.
- Button 16—Deselect.
- Button 17—Deselect.
- Button 18—All buttons should be deselected.
- Button 19—No change at this time.
- Button 20—No change at this time.

Exercise 1: Using TEXTD2 and DIMENS2   153

- Button 21—No change at this time.
- Button 22—No change at this time.
- Button 23—No change at this time.
- Button 24—No change at this time.

The next illustration demonstrates the contents of the Text dialog. Your options, of course, may differ somewhat because the DRWSTD function is controlled by system administrators.

Typical Text dialog.

Text is inserted in the following steps.

1. Select TEXTD2 > TEXT | SIMPLE. Select the insertion point for the word "ISSUE." Note that a new window has appeared at the bottom of the screen. Use the window to input the text: key in *ISSUE* and press <Enter>. Click on the YES button to end text definition. (Multiple lines of text are inserted later in the exercise.)

2. Using the same text settings, insert all the text except the drawing number, title, and heading. Remember to click on the YES button after inserting each new piece of text.

3. To create the drawing number in a different text size, select TEXTD2 > TEXT | SIMPLE. Change the text height (button 2) to *10* and insert a drawing number.

4. To create a heading, select TEXTD2 > TEXT | SIMPLE. Change the font (Button 1); use a sloping font if available. Verify that the height is still set to *10*. The Text Dialog window options default to the last settings used. Insert a heading (e.g., your company name).

5. For the drawing title, *SIDE PLATE*, select TEXTD2 > TEXT | SIMPLE. Change the font (button 1) to the same one employed prior to step 4. Change the text height (button 2) to 6. The font and height used here will also be used for the notes to be created on the drawing later in the exercise.

Now that all the text is in place on the drawing blank you no longer require the insertion points.

1. Rather than individually select all the points you could use a mulit-select option. Select ERASE > NO SHOW, key in *PTD, and press <Enter>. All points will be placed in the no show area.

2. Unfortunately the above procedure will also place the two points you created on the boundary corners in the no show area. To retrieve these two points, select ERASE > NO SHOW and click on the YES button to swap into the no show area. Select the two points for the boundary corners to return them to the normal working area. Click on the YES button again to swap back into the normal working area.

3. The drawing blank is now complete and you can reset to return to the master workspace. Select DETAIL and click on the RESET icon at the top of the screen.

Inserting a Drawing Blank Detail in the Master Workspace

The next procedure in producing the final manufacturing drawing is to insert the drawing blank detail just created in the master workspace. The detail could be inserted in one of the two views already created. However, a better method is to create a new view and insert the detail in the new one. In this way, you can more easily move the drawing blank around the screen relative to the other two views.

1. To create a new view select AUXVIEW > CREATE | NEW BDG. Because this view does not need to be directly related to the two existing views, you can define it by keying in *XY* and pressing <Enter> and then indicating a position for the new view. At this stage the new view's position is not critical because it will be moved in the next step.

Exercise 1: Using TEXTD2 and DIMENS2   155

2. Give the new view a name (i.e., *PLOT*). Save the model.

3. To insert the drawing blank detail into the new view select DETAIL > DITTO | MODEL | STANDARD | SINGLE. If you have only one detail in a model, when you press <Enter> to view the list of available details, CATIA will default to use the single detail. If you have more than one detail in a model, when you press <Enter> the list of available details will display and you can select the required detail. Another method of viewing available details is to click on the YES button. In the resultant display of available choices you can select the desired detail.

4. To insert the selected detail to the current view, *PLOT,* key in , and press <Enter>. The detail is inserted on the new axis.

The next phase is to rearrange the views to resemble the next illustration.

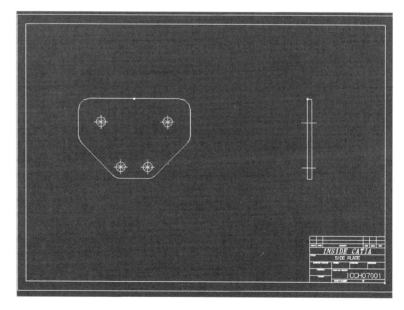

Two views of side plate and drawing blank.

1. Select AUXVIEW > MODIFY | TRANSLAT, and move the views around the screen as described in Chapter 6.

2. Save the model.

Creating a Symbol of Drawing Blank

Because the drawing blank was inserted with the use of the DETAIL function, you can use the detail more than once. The only problem with using this detail of an A1 drawing blank is that all the information is reproduced in every ditto. Naturally, you would want to use a different title and drawing number. A workaround would be to create a symbol of the A1 drawing blank, and then use the symbol in a manner similar to that of a detail. With the use of SYMBOL, you can alter the text on a single occurrence of the symbol without affecting any other occurrence of the symbol.

To experiment with the SYMBOL function as an alternative to using DETAIL, you could delete the occurrence of the A1 drawing blank detail from the model just saved, create a symbol of the A1 drawing blank, and use the symbol instead. Take the following steps.

1. Read the model used in this exercise. Using ERASE, delete the A1 drawing blank detail from the screen.

2. Verify that that you are in the *PLOT* view from which the detail was deleted by using AUXVIEW > CHANGE, or click on the VU button on the fixed menu area.

3. With DETAIL > CHANGE, switch to the detail workspace of the A1 drawing blank.

4. Select SYMBOL > DEFINE | DUPLICATE. Key in the name for the symbol, *A1 SYMBOL*, and press <Enter>.

5. Key in a comment or click on the YES button to continue. Click on the YES button again to confirm creation of the symbol. Click on the RESET button at the top of the screen to return to the master workspace. You have now created a symbol of the A1 drawing blank.

At this stage, the drawing blank symbol will be inserted into the view called *PLOT.*

1. Select SYMBOL > MODEL | STANDARD | SINGLE. If there is a single symbol in a model, CATIA will default to using that symbol when you press the <Enter> key to view the list of available symbols. If multiple symbols exist, you can select the required symbol when the list is

Exercise 1: Using TEXTD2 and DIMENS2

accessed after pressing <Enter>. Another method of viewing available symbols is to click on the YES button. In the resultant display of available choices you can select the desired symbols.

2. To insert the selected symbol to the current view, *PLOT,* key in , and press <Enter>. The symbol is inserted on the current axis.

Now that the symbol has been inserted to the view, you can proceed to changing the text without returning to a detail workspace. Take the following steps.

1. Select SYMBOL > EXTRACT | REPLACE, and then select the symbol.

2. Select the text that you wish to change. Assume that you wish to change the drawing number. The text is now extracted from the symbol and can be modified in the same manner as any other text inserted in the model.

3. Upon selecting TEXTD2 > MODIFY | TEXT | EDIT, a Key Text window appears at the bottom of the screen. Select the drawing number that you have just extracted from the symbol. The drawing number will now appear in the Key Text window.

4. Change the drawing number and press <Enter>. The drawing number on the A1 drawing blank symbol has been changed. (The drawing number in other occurrences of the symbol would not change.)

As demonstrated above, using SYMBOL instead of DETAIL when you require more than a single drawing blank in a model can be very useful.

At this stage of the exercise, you have the following three choices of how to proceed.

- Save the model with the same name as before and overwrite the existing model. In this way you would retain only the model with the drawing blank used as a symbol.

- Save the model with a new name. This option retains the model with the drawing blank used as a detail and as a symbol.

- Reading the existing model without saving the model with the drawing blank used as a symbol. Therefore, you would retain only the model with the drawing blank used as a detail.

Saving the model with a new name (the second choice) is recommended. In this way you would give yourself the chance to return to the model with the drawing blank used as a detail at a later time. If you wished to use the drawing blank in future models you could transfer the detail into the library and use the same detail over and over again. Of course, your options in this regard depend on how your library is controlled.

Dimensioning Views

The next stage in the exercise is to dimension the two views. Before proceeding, dimension settings must be defined. See the next illustration and the following list to set parameters. Again, your options may differ somewhat from those shown in the illustration because the window is typically controlled by system administrators via the DRWSTD function.

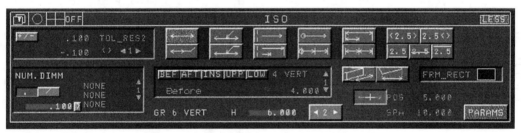

Typical Dimension Dialog window.

- Button 1—Switched off.

- Button 2—Top left button in this group (single dimension line with arrow at both ends).

- Button 3—Switched off.

- Button 4—Switched off.

- Button 5—If available, set numerical display choice to NUM.DIMM, or set to a standard from selection. Set accuracy to 0.1.

- Button 6—Deselected.

- Button 7—Switched off.

- Button 8—Switched off.

Exercise 1: Using TEXTD2 and DIMENS2

- Button 9—Switched off.
- Button 10—Select the text font desired for the dimension text. The list of available fonts is accessed by selecting the font choice already in the box. Text height should be set to 6.00.
- Button 11—Switched off.
- Button 12—No change.
- Button 13—No change.

1. Begin with the side view. There are two dimensions to create: thickness and height. Select DIMENS2 > CREATE | DISTANCE.

2. The first dimension to be created is the plate thickness of 10. Select one side line and then the second side line. Both lines are highlighted. When asked to position the dimension, indicate using mouse button 2 in the desired area. If the dimension line is not in the correct position, select the dimension line. Next, indicate a new position by selecting the dimension line first and then indicating a position; the dimension value will not move relative to the extension line. If you want to move the dimension value, simply indicate a new position without first selecting the dimension line. These modifications are possible only if you undertake the modifications directly after creating the dimension. Additional modifications will be examined later.

3. The next dimension to create is the plate at a height of 172. Select DIMENSION > CREATE | LENGTH. When you select the right side line, it will be highlighted. When asked if you wish to dimension the whole line, click on the YES button. Position the dimension by indicating. Similar to step 2, you could now move the dimension value or line.

4. If you wish to modify the position of the 10 dimension line, select DIMENS2 > MODIFY | DIM LINE. Select the dimension line to be moved and indicate a new position for the dimension line. Click on the YES button when the required modification is complete.

5. If you wish to modify the position of the 10 dimension value, select DIMENS2 > MODIFY | VALUE | LOCATION. Select the dimension line to be moved and indicate a new position for the dimension value.

6. To create the plate width dimension at 248 in the front view, select DIMENS2 > CREATE | DISTANCE. Although the dimension is in a different view, there is no need to switch to that view. As soon as you select a line in the front view it becomes the current view. Select one side line and then the second side line. These lines are now highlighted. When asked to position the dimension, indicate using mouse button 2 in the desired area.

For the next few dimensions, you should change the presentation mode (button 2). Select the cumulative mode button (bottom right in the group of 10). This button shows two dimension lines parallel to each other.

1. The next dimensions to be created are the 50 and 148 dimensions. Select DIMENS2 > CREATE | DISTANCE. Select the left line and then the left hole center line; the two lines are now highlighted. When asked to position the dimension, indicate using mouse button 2 in the desired area. Because you are in the cumulative dimension mode, you need only to select the right hole center line and the 148 dimension will be placed in line with the 50 dimension. Click on the YES button to continue with the next dimensions.

2. The other two groups of cumulative dimensions, 50 and 97, and 44 and 50, can now be created in the same manner as the 50 and 148 dimensions. When you move the positions of these cumulative dimensions, note that the two dimensions (50 and 148) move together.

When using DIMENS2 > MODIFY | DIM LINE to move the positions of dimension lines as well as indicating the new position, you can select an existing dimension in the same view and the dimension line will be placed parallel to the selected dimension line.

The next stage involves creating the radius and diameter dimensions. The drawing has been overdimensioned with hole dimensions in order to illustrate a few available dimensioning methods.

1. The first radius dimensions to be created are the R25 corner radii. Select DIMENS2 > CREATE | RADIUS | CIR CNT. Select the radius to be dimensioned and indicate a position for the dimension. If the dimension is not

Exercise 1: Using TEXTD2 and DIMENS2

in the correct position, you can simply indicate a new position and the dimension will move. If you need to move the dimension at a later time, use the DIMENS2 > MODIFY options. (The TYP text will be added later.)

2. The next dimension to create is the 21 diameter hole at the top right of the plate. Select DIMENS2 > CREATE | DIAMETER | CIR CNT. Select the circle to be dimensioned and indicate a position for the dimension. If you wish to move the dimension, simply indicate a new position now or move it later using the DIMENS2 > MODIFY options.

3. For the 21 diameter hole at the bottom left of the plate, select DIMENS2 > CREATE | DIAMETER | CIR TGT. The Position Management window appears at the bottom left of the screen. The following list provides descriptions of window options.

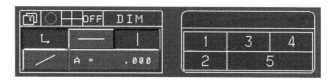

Position Management window.

- Button 1—If this button is selected, the dimension created is positioned relative to the current view.

- Button 2—If this button is selected, the dimension created is positioned relative to a selected element.

- Button 3—If this button is selected, along with button 1, the dimension created is positioned horizontally.

- Button 4—If this button is selected, along with button 1, the dimension created is positioned vertically.

- Button 5—The number entered here defines the angle of the created dimension if button 2 is selected. If zero is entered, the dimension will be parallel to the selected element.

4. In the Position Management window, select buttons 1 and 4 to create a vertical dimension. Select the circle to be dimensioned and indicate a position for the dimension.

5. The last dimension to be created is the 45° angle. Select DIMENS2 > CREATE | ANGLE. Select the bottom horizontal line, and then indicate to the right of the line because the angle dimension is not being created directly between two lines.

6. Select the angled line. Click on the YES button to end selections, and then indicate a position for the dimension. Finally, click on the YES button to end the definition of this dimension.

➥ **NOTE:** *When creating angular dimensions you must select the lines in a counter-clockwise direction.*

To add text to the R25 dimension, take the following steps.

1. Select DIMENS2 > MODIFY | TEXT | ADD. When you select the R25 dimension a Text Management dialog appears at the top of the screen.

Text Management dialog.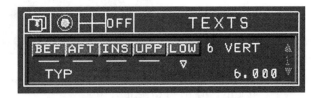

2. From the window select the short line below the word LOW. You can then enter the text that will appear below the selected dimension. Key in *TYP* and press <Enter>.

3. You can also change text height in the same window. Change the height to 6.00.

4. Select the LOW icon; the required text appears below the dimension. To remove the text, select the LOW icon and the text will disappear.

Creating Leaders

To create the leader note, *4 HOLES Ø21*, take the following steps.

1. Select TEXTD2 > CREATE | TEXT | LEADER. Indicate a position for the text to be inserted. A Text Management window appears at the bottom of the screen. Key in *4 HOLES Ø21* and press <Enter>.

Exercise 1: Using TEXTD2 and DIMENS2

2. When the text appears on the screen, it will be underlined, and the numbers 1, 2, and 3 appear on the line. These markers indicate the positions from which the leader line can begin. Select *1*. The Leader Management dialog appears at the top of the screen. The following list describes options in the window which determine leader appearance and the end symbol used.

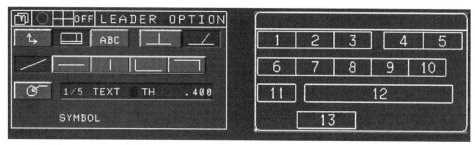

Leader Management dialog.

- Button 1—Along with buttons 7 to 10, leader line will be referenced to the current view.

- Button 2—Along with buttons 7 to 10, leader line will be referenced to the screen.

- Button 3—Along with buttons 7 to 10, leader line will be referenced to the current text.

- Button 4—Leader line will always be normal to the selected element for the anchor point of the end point.

- Button 5—Leader line will always be at a straight angle between start point and element selected for the anchor point of the end point.

- Button 6—Leader line will be straight between the two selected points, or indicated for the anchor points regardless of how buttons 1 to 3 are set.

- Button 7—Leader line will always be horizontal with reference to buttons 1 to 3.

- Button 8—Leader line will always be vertical with reference to buttons 1 to 3.

- Button 9—Leader line will always be horizontal and then vertical with reference to buttons 1 to 3.

- Button 10—Leader line will always be vertical and then horizontal with reference to buttons 1 to 3.

Button 6 is used to define the end symbol. The end symbol can be set differently for indicated positions and selected positions. Selecting the symbol icon will display another management window containing two options that define the symbols used for indicated and selected positions. If these options are selected, a dialog box appears displaying available symbol options. (The options are defined by the system administrator via the DRWSTD function.)

   3. Select buttons 1, 5, and 6, and then indicate a position for the end of the leader. A leader line is created directly from point 1 to the indicated position. Click on the YES button to finish the leader text.

Creating Notes

The next procedure is to create the note. Prior to creating the note you should switch the current view from the front to the plot view. Click on the VU button in the fixed menu area; the current (front) view is highlighted. To switch to the plot view, select any element in that view.

   1. Select TEXTD2 > CREATE | TEXT | SIMPLE. In the Management Dialog window, select the underscore option in button 6.

   2. Indicate a position for the word "NOTE" to be inserted. In the Input text window, key in *NOTE* and press <Enter>. The text appears in the correct position with an underscore. Click on the YES button to complete the entry.

   3. To input the note text, select TEXTD2 > CREATE | TEXT | SIMPLE. In the Management Dialog window, switch off the underscore option by selecting the option in button 6.

   4. Indicate a position for text insertion. In the Input text window, key in *1. REMOVE ALL BURRS AND SHARP EDGES*. Press <Enter>. The text appears in the correct position. Click on the YES button to complete the entry.

   5. Save the model.

Exercise 1: Using TEXTD2 and DIMENS2   165

❖ **NOTE 1:** *If more than a single line of text were required, instead of clicking on the YES button you could select the RETURN arrow option in button 10, and then key in more text. The additional text would have appeared on the next line, and would have been associated with the first line of text. "Association" in this context means that all lines would be moved as a single unit.*

❖ **NOTE 2:** *You can delete text in TEXTD2, without having to select the ERASE function. When using TEXTD2 > DELETE, you have the option of deleting the entire text, a single line of text or a piece of subtext, that is, subscripts or superscripts.*

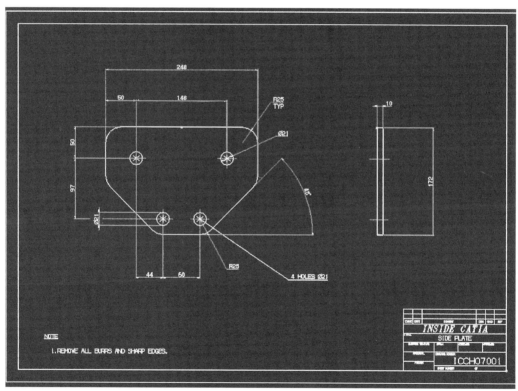

Side plate manufacturing drawing.

Before proceeding to the next exercise, consider experimenting with the TEXTD2 and DIMENS2 functions to view the results of alternative options.

Exercise 2: Using DRAFT and PATTERN

With the exception of PATTERN, all functions and options used in this exercise have been described previously. Therefore, full descriptions of the latter are not included.

1. Read the model saved in Exercise 1 (side plate manufacturing drawing). The first stage of this exercise involves creating a new draft. The reason for creating a new draft is that if only the wheel exercise is showing instead of the side plate and the wheel, manipulation of the screen will be faster and confusion between the two drawings will be avoided.

2. Select DRAFT > CREATE. When asked for a name (in message area), key in *WHEEL* to the input information area and press <Enter>. The new draft is created.

3. To create a new view in which you can produce the front view of the wheel, select AUXVIEW > CREATE | NEW BGD. Because this view does not need to be directly related to the existing views, you can define the view by keying in *XY* and pressing <Enter>. Indicate a position for the new view using AUXVIEW > MODIFY | TRANSLAT. Give the new view a name, such as *WHEEL FRONT.*

Now that the new view is created you can start drawing the front view to the dimensions shown in the target drawing. The following steps are required.

1. Create the circles using CURVE2 > CIRCLE | DIAMETER.

2. Create the center lines using MARKUP > AXIS.

3. Create the top web using LINE > PARALLEL | UNLIM. Create two lines on both sides of the vertical center and then use LIMIT1 > RELIM | TRIM EL1 to relimit the lines to the correct length.

4. Once the top web has been created, the remaining two webs can be created using TRANSFOR > ROTATE.

Now that the first view of the wheel has been created, you could proceed to the creation of the side view section. However, before creating the side view, another look at the DRAFT function is in order.

Exercise 2: Using DRAFT and PATTERN

1. The first step is to delete the duplicated views in the draft called *WHEEL*. Select AUXVIEW > DELETE and then select the three views of Exercise 1. Only one view of the wheel remains on the screen. To see the three views of Exercise 1, which are in the original draft, select DRAFT > CHANGE.

2. You are then asked to click on the YES button to list all available drafts. In this case, there is only one alternative draft. Select **DRAFT* from the resulting list. Once again, the three views of Exercise 1 appear, but not the front view of the wheel. Thus, the three views of Exercise 1 are located in **DRAFT*, and the front view of the wheel is located in the draft called *WHEEL*.

To continue with Exercise 2, take the following steps.

1. Select DRAFT > CHANGE and revert to the *WHEEL* draft.

2. To create the second view select AUXVIEW > CREATE | NEW BGD, and then select one of the vertical lines to define the new view. Click on the YES button to ensure that the new view is correctly associated with the first view.

> **NOTE:** *The views created in the previous exercise were third angle projection; AUXVIEW defaults to the last used choice of view definition. Defining where you require first or third angle projection is not necessary.*

3. Define where the new view will be positioned. Indicate its position to the right of the front view.

4. Name the view *SECTION*.

Continue creating elements in the view by taking the following steps.

1. Create the vertical (to the screen) lines to the dimensions shown in the target drawing using LINE > HORIZONT | UNLIM.

2. Project the horizontal (to the screen) lines across from the front view to the section view using COMBIVU > LINES.

3. Using LIMIT1 > RELIMIT | TRIM ALL, trim corners to provide finished geometry for the section view.

4. To create the angled lines use POINT > PROJ INT to create intersection points, and then use LINE > ANGLE | UNLIM to create the lines.

5. Using LIMIT1 > RELIMIT and LIMIT1 > CORNER, trim the corners to provide finished geometry for the section view.

6. The center line can be created using LINE > VERTICAL | SEGMENT | TWO LIM. Change its graphical representation using GRAPHIC > MODIFY | SAME with the MARKUP center lines in the other view as the reference graphical element.

➥ **NOTE:** *You could create this view by creating only the geometry for a half of the view, and then using TRANSFOR > SYMETRY to create the other half.*

The last stage in creating the section view is to create the hatching pattern.

1. Select PATTERN > AUTO | HATCHING. Click on the YES button to view available hatching patterns. Once again available patterns are controlled by the system administrator via the DRWSTD function. Select a hatching pattern. Note that all solid line and curve elements in the section view are highlighted.

2. To apply the required pattern, indicate in the desired areas. The pattern is inserted in the indicated areas. Click on the YES button to end the selection.

3. In addition to showing where a view has been sectioned, hatching can also be used for analysis. Select ANALYSIS > NUMERIC and then choose the hatching. Note that this selection provides the following information: perimeter center of gravity, cross sectional area, and main moments of inertia.

✓ **TIP:** *If you do not require hatching for anaylsis, consider relegating the hatching into NO-PICK using the ERASE > NO PICK function or the NP button in the fixed menu area. This procedure can facilitate dimensioning because you will not be able to select the hatching by mistake when trying to select a line.*

The following steps are focused on creating an A2 size drawing sheet blank to the sizes shown in the next illustration. The drawing sheet blank will be created as a detail. Because an A1 drawing blank sheet has already been created, you can use the title block from this detail to create the new one.

Exercise 2: Using DRAFT and PATTERN 169

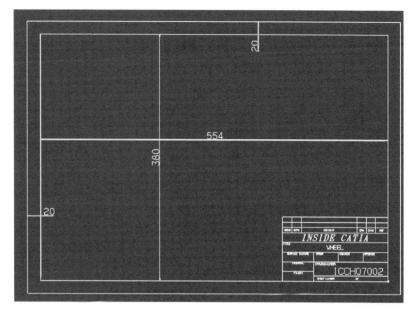

Overall dimensions of A2 blank.

1. Select DETAIL > CREATE. Key in *A2 BLANK* and press <Enter> to name the new detail. Click on the YES button, or key in a comment as desired.

2. With the creation of the new detail, the workspace switches to the detail workspace. Verify the workspace by checking the *W/space : A2 BLANK* message at the bottom right of the screen.

3. To use the title block from the *A1 BLANK* detail, select DETAIL > DITTO | MODEL | STANDARD | SINGLE. With nothing in the input information area, press <Enter> to view a list of the available details. Select *A1 BLANK*. (If only one detail is available, CATIA will default to that detail.) To insert the *A1 BLANK* detail, simply key in , and press <Enter>.

4. Select DETAIL > EXPLODE. Choose the *A1 BLANK* detail; the detail is highlighted. Click on the YES button to confirm that you wish to explode the detail.

5. Because the boundary lines in their current positions are not necessary, select ERASE and delete the lines. Create new boundary lines using LINE > PT-PT | SEGMENT | HOR-VERT and LINE > PT-PT | SEGMENT | VERT-HOR in the same manner as for the *A1 BLANK* detail.

6. Modify the title and drawing number text using TEXTD2 > MODIFY | EDIT.

7. The A2 drawing blank is now complete and you can revert to the master workspace. Select DETAIL, and choose the YES icon at the top of the screen, or click on the WSP button in the fixed menu area followed by clicking on the YES button.

8. Save the model.

The next stage is to insert the newly created drawing blank detail in the master workspace. This detail could be inserted in one of the two views already created. However, a preferred method would be to create a new view and insert the detail in this view. In such fashion, you can more easily move the drawing blank around the screen relative to the other two views.

1. To create a new view, select AUXVIEW > CREATE | NEW BDG. This view does not need to be directly related to the two existing views. Thus, you can define the view by keying in *XY* and pressing <Enter>. Next, indicate a position for the new view. At this stage its position is not critical because it will be moved later.

2. Give the new view a name, such as *PLOT2*.

3. Save the model.

At this point, you should insert the drawing blank detail into the new view and rearrange the views.

1. Select DETAIL > DITTO | MODEL | STANDARD | SINGLE. Upon verifying that nothing is in the input information area, press <Enter>. A list of available details displays; select *A2 BLANK*. An alternate method of viewing details is to click on the YES button.

2. To insert the selected detail in the current view, *PLOT,* key in , and press <Enter>. The detail is inserted on the new axis.

3. The next stage is to rearrange the views as shown in the next illustration. Select AUXVIEW > MODIFY | TRANSLAT, and move the views around the screen as described in Chapter 6.

Exercise 2: Using DRAFT and PATTERN 171

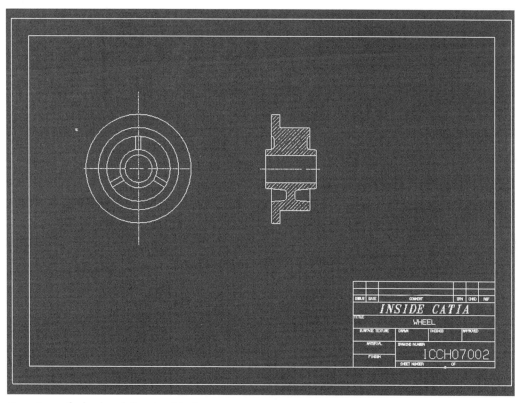

Two views of wheel and drawing blank.

The final stage is annotation.

1. To annotate the drawing, see Exercise 1. Dimension the drawing as shown in the target drawing. Once the dimensioning is completed the text can be inserted.

2. To create the section arrows, place limit points on the vertical center line using POINT > LIMITS, select MARKUP > ARROW | HORIZONT, and use the points as the insertion point for the arrows. The size of the arrow can be set in the MARKUP > ARROW function before creating the arrow.

3. Once the annotation is complete, switch to *DRAFT* to display Exercise 1. Select DRAFT > CHANGE and select *DRAFT*.

4. Save the model.

Before proceeding to the next chapter, experimenting with TEXTD2 and DIMENS2 is recommended. Examples of experimentation follow: (1) Alter TEXTD2 parameters and observe effects on created text; (2) Change the buttons selected in the TEXTD2 management window and observe effects on created text; (3) Alter DIMENS2 parameters and observe effects on created dimensions; and (4) Change the buttons selected in the DIMENS2 management window and observe effects on created dimensions.

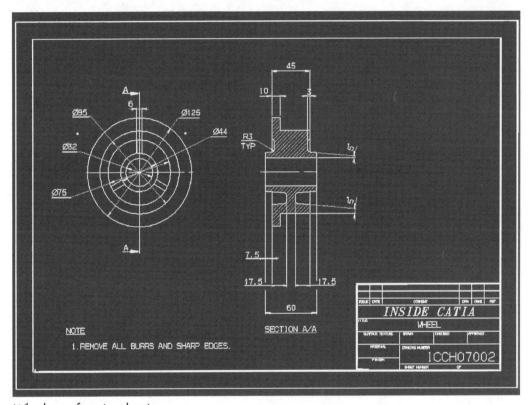

Wheel manufacturing drawing.

Summary

This chapter covered using multiple views to obtain a complete drawing, benefits of using multiple drafts in a model, benefits of using symbols instead of details in models, dimensioning drawings, and placing text and patterns in drawings.

8 Hardcopy Output Using PLOT

The purpose of this chapter is to demonstrate hardcopy production using the PLOT function. Plotting arrangements vary considerably from site to site. Although your system administrator should advise you of the precise procedures, general principles are discussed here.

PLOT

The PLOT function is used to create and manage plot sheets, capture models in plot windows, and arrange plot windows on sheets and preview sheets prior to plotting. A plot sheet contains the plot window(s) stored in a file or profile, and a plot window is the area of the screen that you wish to capture.

Chapter 8: Hardcopy Output Using PLOT

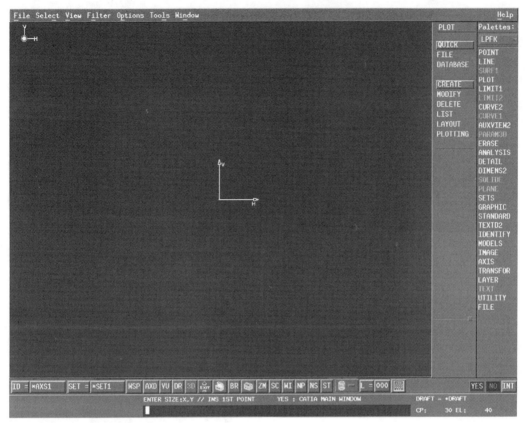

PLOT function in DRAW mode.

At some sites, it is also possible to store plotting parameters in a database; you could then use the database to submit a plot. (Check with your system administrator to determine whether you have access to this facility.) Options available in the PLOT function are described in the next table.

Option	Description
QUICK | CREATE	Create a quick plot without storing it in a sheet file or database.
QUICK | MODIFY	Modify a plot window in quick mode.
QUICK | DELETE	Delete windows from the current sheet, either directly or after displaying them first.

PLOT

Option	Description
QUICK | DELETE | DIRECT	Delete windows from the current sheet directly without displaying them first.
QUICK | DELETE | VISUALTN	Delete windows from the current sheet after displaying them first.
QUICK | LIST	List plot windows.
QUICK | LAYOUT | LOCATE or PARM TEXT	Lay out windows, define sheet format, preview prior to plotting, and manage parameterized texts.
QUICK | PLOTTING | PRINT	Print the plot.
QUICK | PLOTTING | MANAGE | SCALE or FORMAT	Manage the plotting format and scale tables in the project file. (Typically, only system administrators have access to this option.)
QUICK | PLOTTING | PROFILE | BROWSE or DELETE	Browse edit or delete profiles.
FILE | SHEET	Define the current sheet file and the current sheet.
FILE | SHEET | CREATE	Create a sheet in the current sheet file.
FILE | SHEET | READ	Define the current sheet file and the current sheet.
FILE | SHEET | COPY	Copy an existing sheet to create a new sheet or replace an existing sheet in the same sheet file or another sheet file.
FILE | SHEET | MOVE	Move an existing sheet to create a new sheet or replace an existing sheet in another sheet file.
FILE | SHEET | DELETE	Delete sheets from the current sheet file.
FILE | SHEET | MERGE	Merge the windows of two different sheets.
FILE | SHEET | RENAME	Rename a sheet.
FILE | WINDOW	Create a window in the current sheet.
FILE | WINDOW | MODIFY	Modify the current window by translation by modifying its frame or by zooming or applying filters.
FILE | WINDOW | COPY	Copy windows in the current sheet or from an external sheet to the current sheet.
FILE | WINDOW DELETE | DIRECT	Directly delete windows from the current sheet.
FILE | WINDOW | DELETE | VISUALTN	Delete windows from the current sheet by displaying them first.
FILE | WINDOW | LIST	List the plot windows.
(FILE | LAYOUT | LOCATE or PARM TXT	Define or modify the current sheet format, modify windows layout, choose a sheet format, preview the sheet prior to plotting, and define parameterized texts.

Option	Description
FILE | PLOTTING | PRINT	Print the plot.
FILE PLOTTING | PREVIEW	Preview the current or other sheets prior to plotting.

The following options are usually accessible by the system administrator only: FILE | PLOTTING MANAGE; FILE | PLOTTING MANAGE | SCALE; and FILE | PLOTTING | MANAGE | FORMAT. Additional options are available for users who employ profiles and a database. Contact your system administrator for further information.

The following exercises focus on some of the most frequently used PLOT options.

Exercise 1: Creating a Full-size Quick Plot

1. Load the *SIDE PLATE* drawing saved in Chapter 7. Select PLOT > QUICK | CREATE.

2. In the plot information window, ignore the Length and Width settings, and click on the blue bar at the right of Scale. Input *1* and press <Enter>. (Before pressing <Enter>, verify that all digits remaining from a previous entry have been deleted.)

3. Select the X point at the top left of the sheet frame. Select the point at the bottom left corner while dragging the plot window frame across the workspace until it fits the plot sheet. The WINDOW NUMBER 1 CREATED messages appears, as well as the YES : NEW WINDOW prompt.

4. Because only one window is required in this exercise, select PLOT > QUICK | PLOTTING | PRINT.

Exercise 1: Creating a Full-size Quick Plot

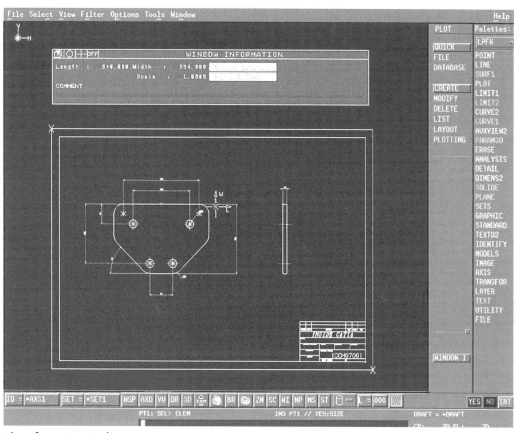

Plot information window.

> **NOTE:** Note that a BR button is displayed at the bottom left of the screen. The button enables buffer regeneration (refreshing the visual graphical representation of the work area if it has become distorted). Buffer regeneration is normally performed by clicking on the BR button in the fixed menu bar, but while you work in the PLOT function, the normal method is unavailable.

178   *Chapter 8: Hardcopy Output Using PLOT*

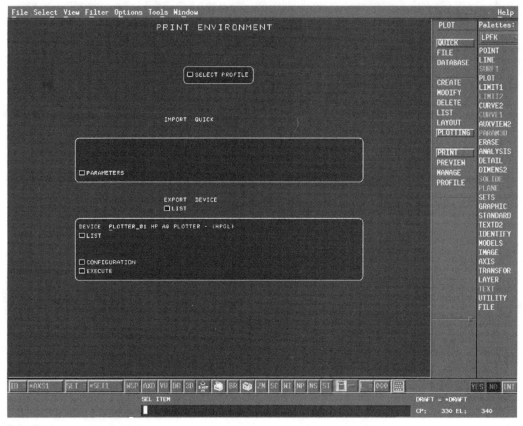

Print Environment screen.

5. The Print Environment screen offers you various options. Selecting SELECT PROFILE activates the Profile Environment window where stored profiles can be read from a database. (Check with your system administrator to determine database availability.)

6. Click on PRINT again in the PLOT menu. Selecting PARAMETERS activates the Sheet Parameters window which contains numerous selection options and entry panels.

Exercise 1: Creating a Full-size Quick Plot

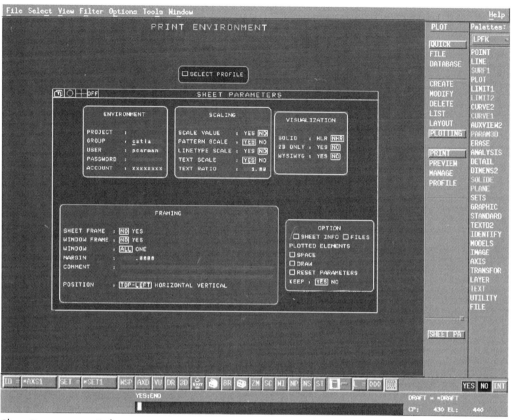

Sheet Parameters window.

> **NOTE:** *You may find that default entries have been set by your system administrator. If not, make selections so that the window matches the above illustration.*

7. Click on the YES button. Ignore the LIST option below the words EXPORT DEVICE. The current plotting device is shown at the top of the next panel. Additional available plotters/printers can be viewed by selecting LIST below the word DEVICE.

8. Select PRINT again from the PLOT menu. When selecting CONFIGURATION, the Output Device Parameter window is activated where selections relating to your particular plotting device can be entered or

modified. Check with your system administrator before making any changes here. Click on the YES button.

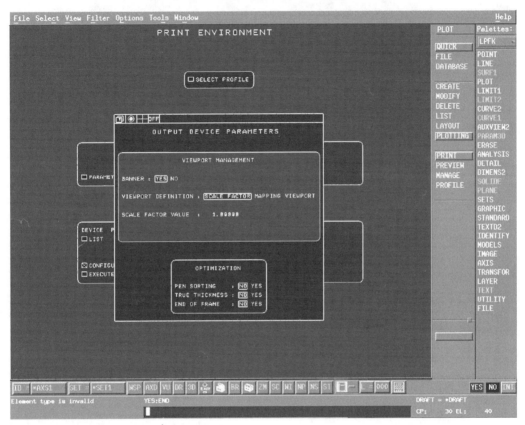

Output Device Parameters window.

9. Upon selecting EXECUTE, the Execution Parameters window is displayed. In the OPERATION MODE panel, select EXECUTE (the first option). The other options related to stored profiles will not be used.

Exercise 1: Creating a Full-size Quick Plot

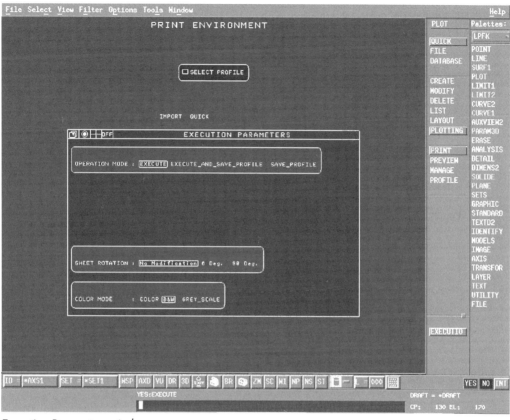

Execution Parameters window.

10. In the SHEET ROTATION panel, select NO_MODIFICATION. This means that the plot window will be sent to your plotter in its present orientation. In the COLOR MODE panel, select B&W for production of a black and white plot. You can also experiment with the other options, COLOR and GREYSCALE. Click on the YES button to execute the plot.

11. The message PROCESS ##### STARTED displays along with the prompt, SEL ITEM, an invitation to make further selections from the panel if required.

12. After a short time a CATIA Submission window appears, providing information about the submitted plot (including whether or not the plot has been successfully executed). The contents of this window may vary depending on your system setup. (Consult your system administrator for further information as necessary.)

Exercise 2: Creating a Multiple Window Plot in Quick Mode

If you do not have the side plate drawing on the screen, read the file now. The objective of this exercise is to produce a plot with three windows: one full size A1 and two half scale A3s.

1. Select PLOT > QUICK | CREATE. Click on the Format Setup button in the Window Information window, and then select A1 from the displayed list.

2. Click on the Scale Setup button in the Window Information window, and then select 100% from the displayed list. Select the top left point, and then the bottom right point. Click on the YES button to create a new window.

3. Click on the Scale Setup button, and then select 50% from the displayed list. Select the top left point, and then the bottom right point. Click on the YES button to create a new window.

4. Select the top left point, and then the bottom right point. Select PLOT > QUICK | LAYOUT. The CURRENT LAYOUT screen is displayed. Click on the AUTOLAYOUT button at the bottom of the screen and then click on the YES button. (You will hear a beep.)

5. In the OPTIONS panel, click on the MOD SHT option. Note that the graphic representation of the plot window layout changes. Click on the ROT SHT button to rotate the plot sheet through 90 degrees.

6. Click on PREVIEW to view the plot sheet.

Exercise 3: Creating and Modifying a Multiple Window Plot

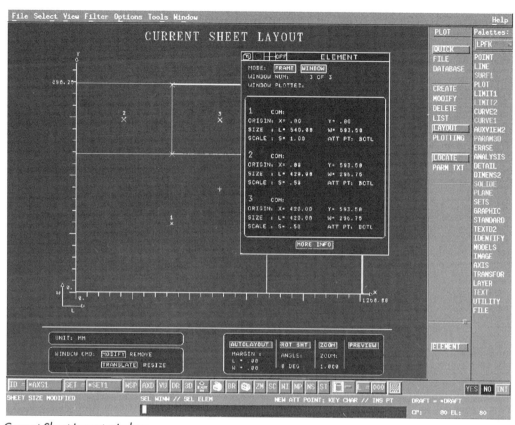

Current Sheet Layout window.

7. Select PLOT > QUICK | PLOTTING, and then EXECUTE. Click on the YES button. After a short time a Catplot Submission window will appear as before. The plot has now been sent to the plotter/printer.

Exercise 3: Creating and Modifying a Multiple Window Plot in FILE Mode

If you do not have the side plate drawing on the screen, read it from the file now. In this exercise, the objective is to produce a plot with multiple windows, and store it in a sheet file. The plot will then be re-read and modified.

1. Select PLOT > FILE | SHEET | CREATE. Select the -> symbol from the right end of the blue bar which represents the file entry area. Next, select a sheet file from the displayed list. (Contact your system administrator about which file sheet to use.)

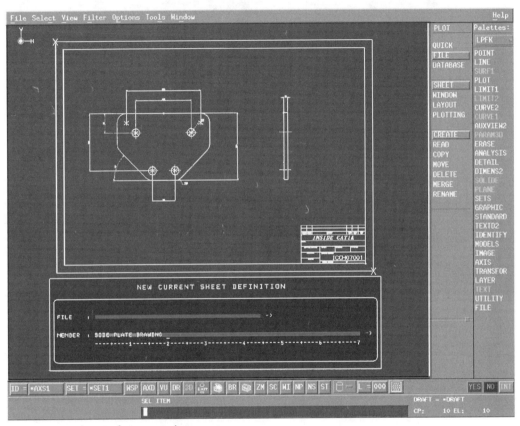

New Current Sheet Definition window.

2. In the MEMBER entry area, type in a name for your plot, or click on the YES button to accept the default model name. Other information windows are displayed, as is the SHEET MEMBER CREATED message.

3. Select PLOT > WINDOW | CREATE. The same procedure as in Exercise 2 can now be followed to create multiple windows. Click on Format Setup and select A1 LANDSCAPE from the displayed list.

Exercise 3: Creating and Modifying a Multiple Window Plot

4. Click on Scale Setup and select 100% from the displayed list. Select the corner points of the plot frame. Click on the YES button for a new window. Click on Scale Setup and select 50% from the displayed list. Select the corner points of the plot frame. Click on the YES button for a new window.

5. Select the corner points of the plot frame. The WINDOW NUMBER 3 CREATED message is displayed. You have now created a three-window plot, and stored it in a sheet file.

The next stage is to re-read the plot sheet from the sheet file for modification.

1. Select PLOT > FILE | SHEET | READ. Select the -> symbol at the right end of the MEMBER entry area in the New Current Sheet Definition window, and then select your stored plot sheet from the displayed list.

2. Select PLOT > FILE | WINDOW | DELETE | DIRECT. In the displayed LIST OF WINDOWS, select the W button of window 3. The SELECTED WINDOW message appears in red next to the selected W button, as well as the prompt, YES : CONFIRM. Click on the YES button (window 3 will be deleted).

3. Select PLOT > FILE | WINDOW | MODIFY. Click on the blue entry area below COMMENT: for window 2; type *This window has been changed to A1* and press <Enter>. Click on the blue SCALE entry area for window 2. Type *1* and press <Enter>.

4. Select PLOT > FILE | LAYOUT. Click on the AUTOLAYOUT button. Click on the YES button, and then click on MOD SHT.

5. Click on ROT SHT to rotate the plot sheet through 90 degrees. Click on PREVIEW to view the plot sheet.

6. Select PLOT > FILE | PLOTTING. Select EXECUTE, and then the YES button. After a short time a Catplot Submission window will appear as before.

Other PLOT options will be discussed in subsequent chapters.

Exercise 4: Producing a Screen Capture

During a CATIA session you may want to capture the precise contents of the screen, such as a drawing and the alphanumeric analysis window (see Chapter 4). CATIA's screen grab facility permits you to capture a complete CATIA screen, including palettes and so forth.

 1. From the pull-down menu bar, select TOOLS > SCREEN GRAB. The following prompt displays: <ENTER> SIZE : X,Y // INS 1st POINT YES : MAIN WINDOW. Click on the YES button. The Screen Grab window appears.

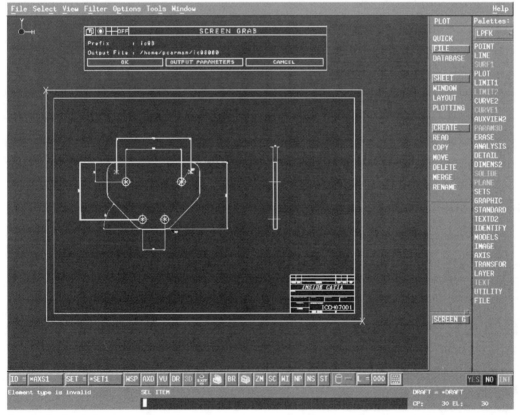

Screen Grab window.

Exercise 4: Producing a Screen Capture

2. In the Screen Grab window, click on the OUTPUT PARAMETERS button. The Screen Grab : Output Parameters window displays. Select Tiff from the IMAGE TYPE options. (You may have other options available depending on your setup.)

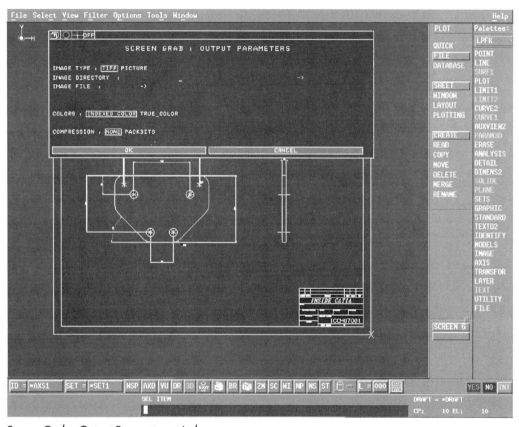

Screen Grab : Output Parameters window.

3. At the right end of the IMAGE DIRECTORY entry area, select the -> symbol, and choose a file from the list in the Metafile window. Click on the IMAGE FILE entry bar. Input a name for the capture file, and press <Enter>.

4. Select either INDEXED COLOR (for eight bits per pixel) or TRUE COLOR (for 24 bits per pixel).

5. Select either None (for no compression) or PACKBITS (to compress to Packbits mode). Click on the OK button, and the Tiff file is created.

It is also possible to create a Tiff file transparently (i.e., without making any menu selections or changing functions) using the /GRABP command. By entering /GRABP in the data entry area, the entire screen is captured based on the last used settings and the last Tiff file name is incremented by one. This method enables you to capture certain windows and panels which are unavailable under the SCREEN GRAB option.

> **NOTE:** *The method used to send the capture to a plotter or printer will vary depending on the setup and the facilities available at your site.*

To capture only part of the screen work area, take the following steps.

1. Select TOOLS > SCREEN GRAB. Indicate (using mouse button 2) the top left corner of a rectangle which will enclose the area to be captured. As you move the cursor to the position of the bottom right corner of the area to be captured, note that a stretchable box is formed. This is called "rubberbanding," and is a feature of several CATIA functions.

2. Alternatively, indicate the center point of the area to be captured, and then type in the length and width of the area as X,Y and press <Enter>. Once again the Tiff file is created and is ready to be sent to a plotter or printer.

Summary

This chapter covered how to create a full size quick plot of a drawing; how to create a multiple window plot in quick mode; how to create and modify a multiple window plot in file mode; and how to create a screen capture.

9 Generating and Modifying SPACE Graphic Elements

The purpose of this chapter is to introduce the SPACE environment. The same functions discussed in Chapters 2 and 3 will be covered here, but working in SPACE mode rather than 2D. Because generating SPACE elements in CATIA is very similar to generating elements in DRAW, a large portion of the material covered in this chapter should be familiar. Major topics covered in the chapter are summarized below.

- Creating geometry in SPACE mode
- Erasing unwanted information
- Relimiting SPACE geometry
- Modifying graphical representation of SPACE mode elements

The only new function introduced in the chapter is PLANE. Additional functions previously introduced in earlier chapters but discussed in SPACE mode include POINT, LINE, CURVE2, ERASE, LIMIT1, and GRAPHIC.

There are two ways of working in SPACE mode: 2D SPACE and 3D SPACE. Two-dimensional SPACE mode is used when working on a plane in three-dimensional space. In contrast, three-dimensional SPACE mode is used when working on non-planar objects in three-dimensional space.

Certain 3D SPACE functions are not available when working in 2D SPACE, whereas different options are available for other SPACE functions depending on whether the user is working in 2D or 3D SPACE mode.

POINT

The POINT function is the same in 2D SPACE mode and 3D SPACE mode. The following table describe POINT options in SPACE mode.

Option	Description
PROJ INT	Select intersecting elements.
PROJECT	Project existing points to other elements.
COORD	Enter X, Y, and Z coordinates.
LIMITS	Place limit points on lines or curves.
SPACES	Place equidistant points on lines or curves.
TANGENT	Place tangency points on planar curves with respect to other elements.

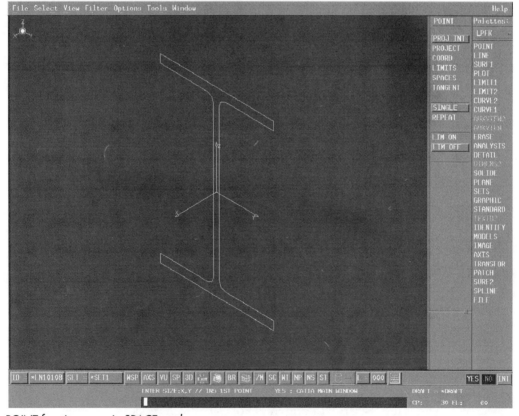

POINT function menu in SPACE mode.

LINE

The LINE function in 3D SPACE is used to create and modify unlimited lines and line segments. Options are described in the next table.

LINE in 3D SPACE Mode

Option	Description
PT-PT	Select points and ends of existing lines or curves.
PARALLEL	Use existing lines to create parallel lines.
NORMAL	Use existing elements to create normal lines.
INTERSEC	Use two planes to create an intersecting line.
PROJECT	Use an existing line and plane to create a new line projected to a plane.
ANGLE	Use points, lines, and planes to create angled lines.
COMPON	Use components of a line.
EDGE	Use point and window plane to create a normal line.
TANGENT	Use points, lines, or curves to create tangential lines.
MEAN	Use at least two points to create a mean line.
POL EDGE	Use the edges of a solid or polyhedral surface to create a superimposed line on the edge.

Lines can also be modified using the MODIFY option from the LINE function menu.

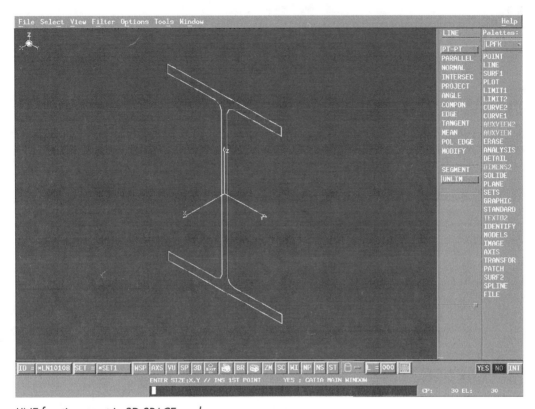

LINE function menu in 3D SPACE mode.

The LINE function in 2D SPACE mode is used to create and modify unlimited lines and line segments. Options are described below.

LINE in 2D SPACE Mode

Option	Description
PT-PT	Selecting points and ends of existing lines or curves.
PARALLEL	Use existing lines to create parallel lines.
HORIZONT	Use points to create horizontal lines.
VERTICAL	Use points to create vertical lines.
NORMAL	Use points and lines to create normal lines.
MEDIAN	Use points or a line to create a median line.
BISECT	Bisecting existing lines.

Option	Description
ANGLE	Use points or lines to create angled lines.
COMPON	Use components of a line.
TANGENT	Use points, lines, or curves to create tangential lines.
MEAN	Use at least two points to create a mean line.
EDGE	Use a point to create a line normal to the current 2D plane.

Lines in 2D SPACE mode can also be modified using the MODIFY option from the LINE function menu.

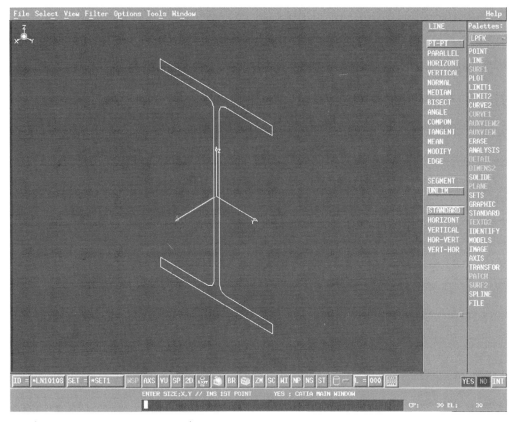

LINE function menu in 2D SPACE mode.

CURVE2

The CURVE2 function in 3D SPACE is used to create and modify curves (circles, conics, and splines). Options are described in the next table.

CURVE2 in 3D SPACE Mode

Option	Description
PARALLEL	Use existing curves to create parallel curves.
CONNECT	Use existing curves to create a connecting curve between the selected curves.
PTS CST	Create curves passing through multiple selected points.
APPROXIM	Modify the degree and tolerance value for a given curve.
TGT CONT	Use tangent continuity at arc limits to create a curve.
CVT CONT	Use curvature continuity at arc limits to create a curve.
DEPTH	Create wireframe volume by transformation, translation, or rotation of a contour.
CIRCLE	Use axis and point or three points to create circles.
HELIX	Define a helix of revolution.
SPINE	Use point on a plane and normal to one or more other planes to create a spine.
INVERT	Invert lines, canonical forms, or curve parameterization.

Circles can also be modified using the CIRCLE | MODIFY option from the CURVE2 function menu.

CURVE2

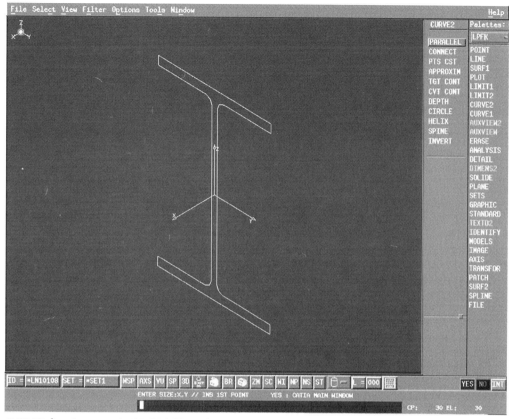

CURVE2 function menu in 3D SPACE mode.

The CURVE2 function in 2D SPACE is used to create and modify curves (circles, conics, and splines). Options appear below.

CURVE2 in 2D SPACE Mode

Option	Description
RADIUS	Define center and radius to create a circle.
DIAMETER	Define center and diameter to create a circle.
THREE-PT	Define three points or two points and radius to create a circle.
PART-ARC	Define three points or two points and radius to create an arc.
MONO-TGT	Use one or more elements to create tangent circles where the center is known or not known.
MULTI-TGT	Use one or more elements to create tangent circles where the center point is known or not known.

Chapter 9: SPACE Graphic Elements

Option	Description
ELLIPSE	Define geometry to create ellipse.
CONIC	Use three or five points to create conic curves.
PTS CST	Create curves passing through selected points.
PARALLEL	Use existing curve to create a parallel curve.
APPROXIM	Modify the degree and tolerance value for a given curve.
TGT CONT	Use the tangent continuity at the arc limits to create a curve.
CVT CONT	Use the curvature continuity at the arc limits to create a curve.
INVERT	Invert lines, canonical forms, or curve parameterization.

Circles can also be modified using the CIRCLE-MODIFY option from the CURVE2 function menu.

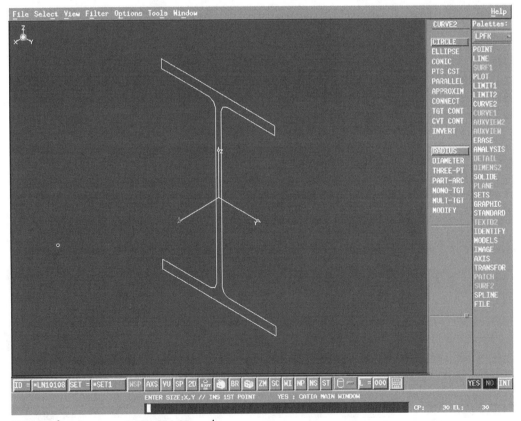

CURVE2 function menu in 2D SPACE mode.

ERASE

The ERASE function is the same in 2D SPACE mode and 3D SPACE mode. Options for the ERASE function in 3D SPACE mode are described in the next table.

Option	Description
ERASE	Delete elements from the model.
PACK	Pack the model after element deletion.
NO SHOW / SHOW	Hide elements from view and retrieve hidden elements.
NO PICK / PICK	Make elements nonselectable and retrieve nonselectable elements.

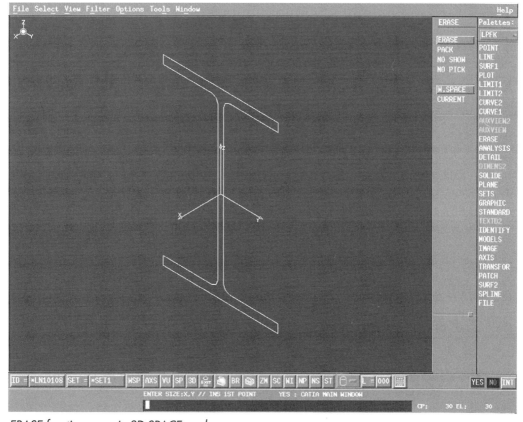

ERASE function menu in 3D SPACE mode.

LIMIT1

The LIMIT1 function is the same in 2D SPACE mode and 3D SPACE mode. The LIMIT1 function is used to modify lines, circles, conics and curves, and can also be used to create connecting elements (e.g., chamfers and fillets). Options are described below.

Option	Description
RELIMIT	Use existing lines, curves, and points to relimit existing lines or curves.
CORNER	Add connecting curves to existing lines and curves.
MACHINE	Add chamfers to existing lines and curves.
BREAK	Break lines and curves.
CONCATEN	Join two lines or curves to create a single element.
EXTRAPOL	Extend the length of a line or curve.

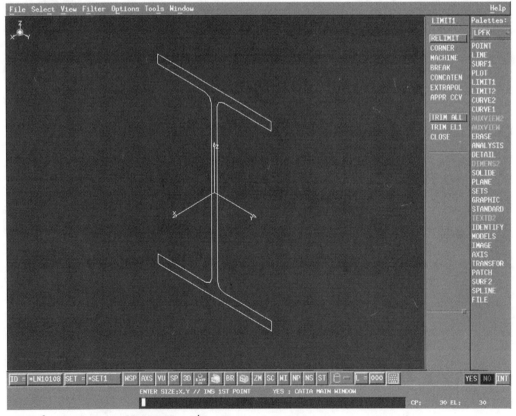

LIMIT1 function menu in 3D SPACE mode.

GRAPHIC

The GRAPHIC function is used to modify the graphical representation of lines, curves, points, solids, faces, and so on. This function is the same in 2D SPACE mode and 3D SPACE mode. The graphical attributes used for any element can also be checked using the ANALYZE option from the GRAPHIC function menu. Options are described in the next table.

Option	Description
LINETYPE	Change line type to solid, dotted, and so on.
PT TYPE	Change point type to star, cross, and so on.
THKNESS	Change line thickness.
COLOR	Change color.
BLINK / STEADY	Highlight elements.
MOD VISU	Change display parameters such as discretization, transparency, and hidden line for solids.
VERIFY	Verify display attributes of an element.

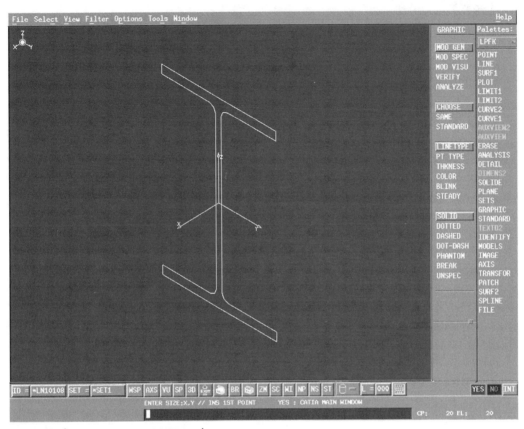

GRAPHIC function menu in SPACE mode.

PLANE

The PLANE function is only available in SPACE mode and is used to create and modify planes. Options are described in the following table.

Option	Description
THROUGH	Use points, lines, curves, surfaces, faces, and functional surfaces of solids to create planes.
EQUATION	Use equation to create plane.

PLANE

Option	Description
PARALLEL	Use existing planes to create parallel planes.
NORMAL	Use existing elements to create planes normal to the selected elements.
ANGLE	Use existing line and plane to create angular planes.
ORIENTN	Modify orientation of an existing plane.
MEAN	Use at least three points to create a mean plane.
PRL WIN	Use point to create a plane parallel to the current screen window.
EDGES	Use points and lines to create a plane normal to the current screen window.
SPACES	Use points, lines, and curves to create equidistant planes.
LIMITS	Use lines and curves to create normal planes at end points.

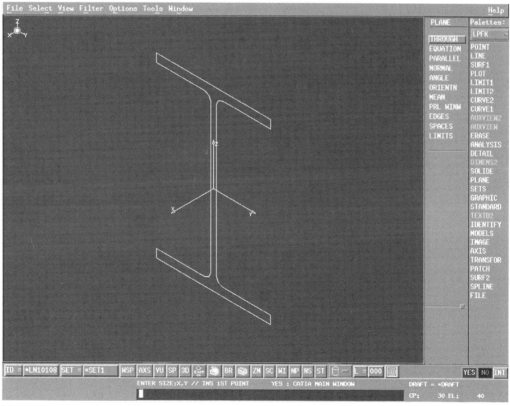

PLANE function in 3D SPACE mode.

Generating and Modifying SPACE Graphic Elements

Exercises in this chapter demonstrate some of the most commonly used options in the seven SPACE mode functions reviewed here. The geometry used in the following exercises was developed in previous chapters, but this time the geometry will be created in space.

Setup for Exercises

In previous exercises, new models were created for working in DRAW mode. For the following exercises, new models must be set up for working in SPACE mode. To create a suitable new model, use the following steps.

1. Select FILE > CREATE | YZ. Before creating the new model, you must be working in SPACE mode. Verify that the SP/DR button on the fixed menu bar reads SP. If the button reads DR, click on the button to toggle to SP.

2. YES:CONFIRM appears in the message area. Click on the YES button; the new model is created, ready for work in SPACE mode. A 3D axis (X, Y, Z) appears in the middle of the work area.

3D and 2D SPACE

Before proceeding to the exercises, a brief discussion of working in 3D versus 2D in SPACE mode is in order. The new model created above is set for 3D SPACE work. To work in a 2D SPACE environment, you would click on the 3D/2D button on the fixed menu bar. To access the 2D option, a plane must be defined. To define a plane, you have the following options.

1. Select an existing plane.

2. Choose existing elements (e.g., three points, two lines, curves, edges of solids, or functional surfaces of solids).

3. Select the X, Y, or Z axis.

4. Key in the plane component in the X, Y, or Z direction.

Generating and Modifying SPACE Graphic Elements 203

In this instance, you can use only the third or fourth options because there is no geometry other than the axis. Using the third option, take the following steps to access the 2D plane.

 1. Click on the 2D/3D button to access the 2D option, and select the X axis. The X axis is dim and no longer selectable. You are now ready to work on the YZ plane. When using LINE, POINT, and CURVE2 functions, working on this plane is very similar to working in DRAW mode. The only difference is that you are working with the 2D plane at an angle to the screen.

 2. To change the angle, click on the WI button on the fixed menu bar, key in *P*, and press <Enter>. The YZ axis is now parallel to the screen. Working in 2D SPACE mode becomes exactly the same as working in DRAW mode.

 3. To switch back to the angled position—normally referred to as the XYZ screen—, click on the WI button, key in *XYZ*, and press <Enter>.

 4. If you wish to revert to 3D SPACE mode, click on the 2D/3D button, and then click on the YES button to reset to 3D SPACE mode.

In the previous example, the third option was used to define the 2D plane. Take the following steps to experiment with using the fourth option.

 1. Click on the 3D/2D button to access the 2D option.

 2. Because in the previous example you used the X axis to define the plane, X=0.00000 appears in the input information area. This is the plane component for the YZ plane. Simply press the <Enter> key, and the 2D plane on which you wish to work is defined.

 3. Click on the WI button, key in *P*, and press <Enter>. You are ready to work with the YZ plane parallel to the screen.

 4. Click on the WI button, key in *XYZ*, and press <Enter>. The screen has reverted to the XYZ screen.

 5. Click on the 2D/3D button, and press on the YES key to return to 3D SPACE mode.

Exercise 1: Creating I Beam Geometry in SPACE Mode

This exercise is a reenactment of the I beam drawn in an earlier chapter. The I beam model created here will also be used in Chapter 11 to create a solid. No dimensions are shown in the next illustration because the required dimensions can be found in Chapter 3, Exercise 6.

200 x 133 universal I beam.

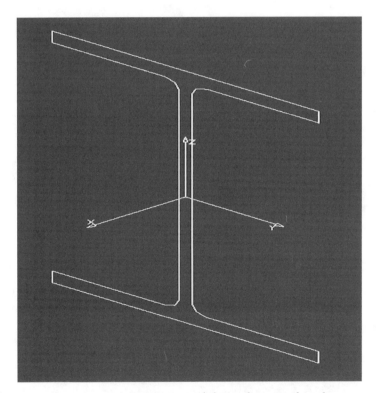

The first stage of this exercise is to create a new model. As discussed in the previous setup section, you should select FILE > CREATE | YZ. Verify that the SP/DR button reads SP; toggle from DR to SP if necessary. Click on the YES button to confirm creation of a new model.

You have two options for how you want to create the geometry in this exercise. First, you can set the screen as seen in the previous illustration. Second, you can

Exercise 1: Creating I Beam Geometry in SPACE Mode

set the screen as demonstrated in the next illustration, that is, view the geometry normal to the plane on which it is to be created.

I beam viewed normal to plane on which it is created.

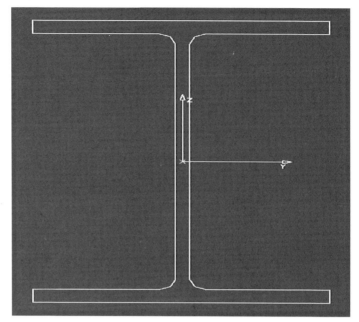

The following steps to create the required geometry were developed on the first option.

1. Define the plane on which the geometry lies, which is also the plane on which you work. Click on the 2D/3D button to change the button display from 3D to 2D. Select the X axis; the X axis dims and becomes nonselectable.

2. Using the same procedure employed in Chapter 3, Exercise 6, create the geometry. For the horizontal lines, use LINE > HORIZONTAL | UNLIM.

3. Create all vertical lines using LINE > VERTICAL | UNLIM.

4. Relimit all corners using LIMIT1 > RELIMIT | TRIM ALL or LIMIT1 > CORNER | TRIM ALL. Remember the LIMIT1 > BREAK function when and if you need to use a line twice.

5. Having created the geometry in 2D SPACE mode, you can return to 3D SPACE mode. Click on the 2D/3D button and then click on the YES button to revert to 3D SPACE mode.

6. Save the model. This model will be used to create a wireframe in Exercise 4 and a solid model in Chapter 11.

The following steps to create the required geometry were developed on the second option of screen settings (I beam viewed normal to plane on which it is created).

1. Create a new model as described above.

2. Define the plane on which the geometry lies, which is also the plane on which you need to work. Click on the 2D/3D button to switch from 3D to 2D. Select the X axis; the X axis dims and becomes nonselectable.

3. To change the display, click on the WI button on the fixed menu area, key in *P* and press <Enter>. The Y and Z axes are now displayed in the same manner as the V and H axes in DRAW mode.

4. Using the same procedure employed in Chapter 3, Exercise 6, create the geometry. For the horizontal lines, use LINE > HORIZONTAL | UNLIM.

5. Create all vertical lines using LINE > VERTICAL | UNLIM.

6. Relimit all corners using LIMIT1 > RELIMIT | TRIM ALL or LIMIT1 > CORNER | TRIM ALL. Remember the LIMIT1 > BREAK function when and if you need to use a line twice.

7. Having created all the geometry in 2D SPACE mode and with the screen display normal to the creation plane, you now need to revert back to the XYZ display. Click on the WI button, key in *XYZ*, and press <Enter>.

8. Having created the geometry in 2D SPACE mode, you can return to 3D SPACE mode. Click on the 2D/3D button and then click on the YES button to revert to 3D SPACE mode. The model is a duplicate of the one created previously.

There is no need to save this model becomes it duplicates the model created in the previous six steps.

Exercise 2: Creating Side Plate Geometry in SPACE Mode

This exercise is a reenactment of the side plate drawn in an earlier chapter. The I beam model created here will also be used in Chapter 11 to create a solid. No dimensions are shown in the next illustration because all required dimensions can be found in Chapter 3, Exercise 7.

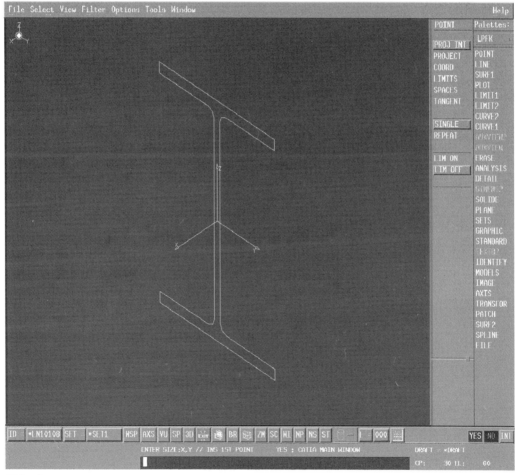

Side plate.

1. Create a new model as described above. Refer to Exercise 7, Chapter 3, for plate dimensions.

2. Create the geometry for the side plate on your own. See Chapter 3, Exercise 7, for guidance if necessary. Remember that the geometry will be created on the YZ plane. Switch to 2D SPACE mode using the 3D/2D button as described earlier in this chapter. Experiment with creating the geometry with the screen display in the XYZ position and normal to the plane on which the geometry is being created. By using both of these options you will be able to decide on the display mode you prefer to work in.

3. Save the model.

Exercise 3: Using POINT and GRAPHIC

Once again, this exercise duplicates a previous model creation exercise. The results of the current exercise will also be used in Chapter 11 to create a solid. No dimensions are shown in the next illustration because all required dimensions can be found in Chapter 7, Exercise 2.

Exercise 3: Using POINT and GRAPHIC

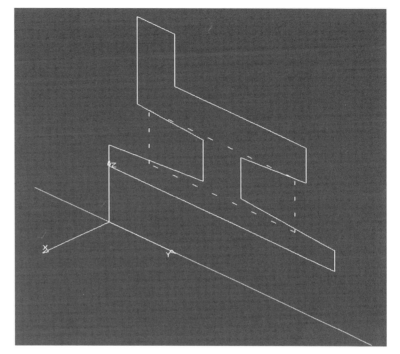

Wheel section.

In contrast to the previous two exercises, you need only create a half of the geometry. The half will be used to create a solid revolution in Chapter 11. To complete this exercise, you will need to create a new model as described above, and refer to Chapter 7, Exercise 2 for wheel dimensions.

Try to create the geometry for this exercise on your own. (Refer to Chapter 7, Exercise 2 for assistance as necessary.) Remember that the geometry will be created on the YZ plane. Switch to 2D SPACE mode using the 3D/2D button as described earlier in this chapter. Experiment with creating the geometry with the screen display in the XYZ position and normal to the plane on which the geometry is being created. By using both of these options you will be able to decide on the display mode you prefer to work in.

When creating the geometry for this exercise do not include the radius corners. You will be able to create them in a different manner when working with solids in Chapter 11.

Another difference compared to previous exercises is that the drawing is created as a section through the recess and not through the web. The dotted lines are created using the web dimensions, and the lower dotted line is created 1mm below the intersection of the vertical dotted and lower angled lines. The reason for this will be explained in Chapter 11. To create the lower dotted line, take the following steps.

1. Create the two vertical dotted lines as unlimited lines using LINE > PARALLEL | UNLIM.

2. Create a point on the intersection of one of the vertical lines just created and one of the lower angled lines using POINT > PROJ/INT.

3. Create a horizontal line through the point using LINE > HORIZONT | UNLIM.

4. Create a parallel line 1mm below the horizontal line using LINE > PARALLEL | UNLIM. Relimit the lines using LIMIT1 > RELIMIT.

5. Change the graphical representation of the lines to dotted using GRAPHIC > MOD GEN | CHOOSE | LINTYPE | DOTTED.

6. Save the model.

Exercise 4: Create Wireframe Model

The objective of this exercise is to create a wireframe model of the I beam section drawn in Exercise 1. The I beam geometry will be duplicated at a distance of 250mm from the original I beam geometry. (The geometry could also be duplicated using the TRANSFOR function. The TRANSFOR function in SPACE mode will be covered in Chapter 13.)

1. Read the model saved in Exercise 1.

2. Upon creation of the original geometry, YZ was defined as the plane on which to work by clicking on the 3D/2D button and selecting the X axis. To create the new geometry, a plane 250mm (9.85") from the YZ plane must be defined. Two methods of defining the plane are described here.

   First, click on the 3D/2D button, key in *X=-250* to the input information area, and press <Enter>. Recreate the geometry on the plane in exactly

Exercise 4: Create Wireframe Model

the same manner as the original geometry was created in Exercise 1.

Second, create a plane 250mm from the YZ plane. Select PLANE > THROUGH, and then select any two lines from the geometry of the I beam on the YZ plane. A highlighted plane appears on the axis; click on the YES button to accept the position of the plane.

> **NOTE:** *Instead of selecting two lines, you could have selected one of the corner curves. A single curve will define a plane because it can lie only on one plane. In contrast, because a single line can lie on an infinite number of planes, a second line is necessary to define the plane.*

3. Select PLANE > PARALLEL and then select the first plane created. A vector normal to the selected plane appears. In this instance, the vector is pointing in the wrong direction. Select the vector to change its direction; key in *250* and press <Enter>. The new plane will be created 250mm away from the first plane.

4. Once the new plane is defined, you can use it to define the 2D plane on which to create the repeated geometry. Click on the 3D/2D button and then select the plane created 250mm from the YZ plane. You could now proceed and recreate the geometry on this plane in exactly the same manner as the original geometry was created in Exercise 1.

5. Return to 3D SPACE mode by clicking on the 2D/3D button. Click on the YES button.

6. Because you no longer require the planes created, you can either erase or hide them using the ERASE function.

At this point, there are two sets of I beam geometry at a distance of 250mm from one another. The next step in completing the wireframe is to create the joining lines.

1. Select LINE > PT-PT | SEGMENT, and join the lines as shown in the next illustration.

> **NOTE:** *When using LINE > PT-PT, selecting the line end to define the end point is not necessary. Select the line at any point less than halfway*

from the end you wish to define; the end point will be selected. Selection of the line on the pertinent half for the end point definition is called "vicinity selection."

2. Save the model with a different name to avoid overwriting the original. The original model will be employed in exercises in Chapter 11.

Wireframe I beam.

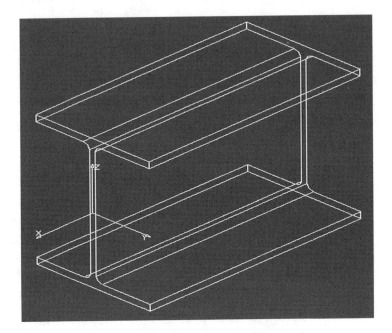

An alternative (and easier) method of producing the wireframe in this exercise would be using the CURVE2 function as described below.

1. Read the model saved in Exercise 1.

2. Select CURVE2 > DEPTH | TRANSLAT. Choose each of the elements in turn or click on the YES button to end selection.

3. Select the X axis, to define the direction of translation, key in *-250* to the input information area, and press <Enter>. The wireframe will then be created automatically.

4. Save the model with a different name to avoid overwriting the original. The original model will be employed in Chapter 11.

Summary

The following topics were covered in this chapter: creating geometry in SPACE mode, modifying elements in SPACE mode, working in 2D and 3D SPACE modes, creating 3D wireframe models, and creating planes.

Before proceeding to the next chapter, consider experimenting with working in 2D and 3D SPACE modes to ensure that you are fully conversant with 2D versus 3D space. Examples of practice runs follow: (1) Use the last exercise to create planes using the available geometry (e.g., a plane on the side of the I beam); and (2) Try switching to different 2D planes, such as in using the lines that form the side of the I beam.

10 Analyzing 3D Graphic Elements

This chapter is focused on analyzing the numerical values of SPACE type elements such as lines, points, curves, and so forth with the ANALYSIS function. Chapter 4 covered analysis aimed at the 2D DRAW mode only. This chapter covers selected uses of ANALYSIS in SPACE mode. Appearing below are ANALYSIS options and respective descriptions.

ANALYSIS in SPACE Mode

Option	Description
NUMERIC	Analyze numerical values of SPACE type elements.
RELATIVE	Analyze relative values of SPACE type elements.
LOGICAL	Analyze logical links between SPACE elements.
CURVE	Analyze variations in curvature and gradient of CCV type elements.
INERTIA	Analyze inertial values of faces, FAC, volumes, VOL and solids, and SOL to provide inertia, area, volume, and mass information.

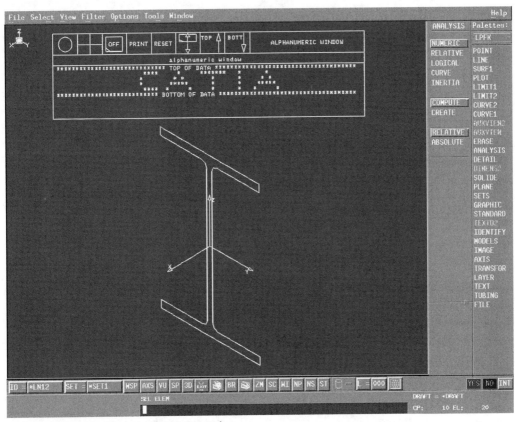

ANALYSIS function menu in 3D SPACE mode.

> **NOTE:** When selecting ANALYSIS in 2D SPACE mode the options available are the same with the exception of INERTIA, which is not available because solids and other 3D objects are not selectable when in 2D SPACE mode.

Exercise 1: Using the ANALYSIS Function

1. Using FILE > READ, load the I beam model prepared in Exercise 1, Chapter 9.

2. Select ANALYSIS > NUMERIC | COMPUTE | ABSOLUTE. The prompt in the message area will tell you to select an element (SEL ELEM).

Exercise 1: Using the ANALYSIS Function

3. Select the top horizontal line of the I beam geometry. The following illustration shows the following items that you should be able to view in the alphanumeric window: CATIA identity for the selected element; start and endpoints of the line; components for the direction vector of the line; angular characteristics of the line relative to the X, Y, and Z axis; and length of line.

➥ **NOTE:** *Selection of the ABSOLUTE option for this exercise means that the analysis is relative to the absolute axis system. In this instance, it is the only axis system.*

If more than one axis system is in use in the model, the current axis system could be used for analysis purposes by selecting the RELATIVE option.

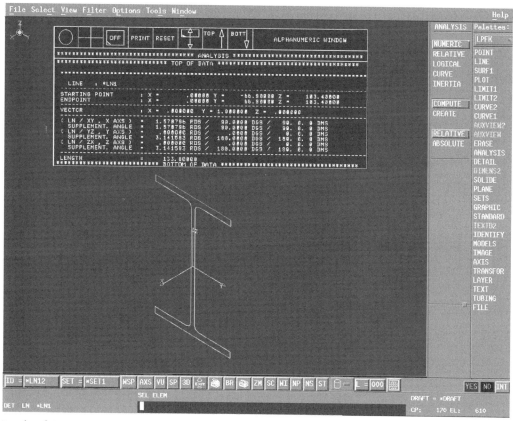

Results of ANALYSIS > NUMERIC.

✓ **TIP:** *If the window does not appear as seen in the previous illustration, click on the circle in the top left corner of the window and experiment with the various controls (i.e., PRINT, RESET, etc.).*

Now that the alphanumeric window is the active window, you can drag it using the method described in Chapter 1. This window can be switched on and off manually by pressing <Alt> and the plus sign <+> key to switch on, and <Alt> and <-> to switch off. The window will automatically switch off when another CATIA function is selected. If you want to manually make the work area active again, select the red circle on the axis icon at the top left corner of the screen.

To perform relative analysis, you could choose any two geometric elements, but for the sake of this exercise, determine the relationship between the top and bottom of the I beam by making the selections in the following steps.

1. Select ANALYSIS > RELATIVE | SINGLE | ABSOLUTE and then select the top and bottom horizontal lines of the I beam drawing. The following information will display in the alphanumeric window: CATIA identities for the selected elements; distance between elements; distance between elements relative to the X, Y, and Z axes; closest elements relative to the X, Y, and Z axes; and confirmation that the elements are parallel to one another.

Exercise 1: Using the ANALYSIS Function

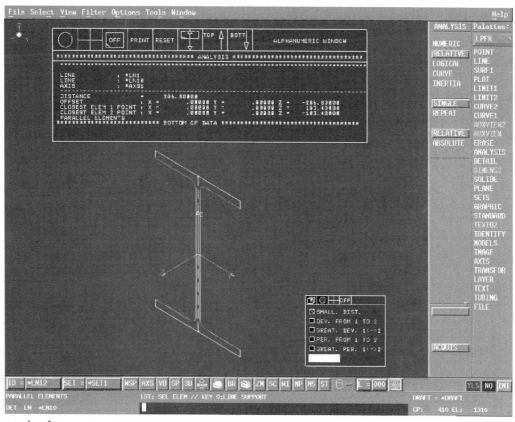

Results of ANALYSIS > RELATIVE.

2. At the bottom of the screen is another window providing additional options: smallest distance (in this case selected); deviation from the smallest distance if the lines are staggered or angled to each other; greatest deviation from the smallest distance if the lines are staggered or angled to each other; perpendicular distance between the selected elements; and greatest perpendicular distance between the selected elements. The option to create permanent elements to replace the temporary elements is shown during analysis. Having performed an analysis, select the CREATE icon and the elements will be created after clicking the YES button.

3. If you required the relative analysis between the top horizontal line and various other elements of the I beam, you do not have to continue selecting the top line and other required elements. Relative analysis between repeated elements can be achieved by selecting ANALYSIS > RELATIVE | REPEAT | ABSOLUTE. If you now select the top element and the bottom horizontal line, the same results as in the previous example appear.

4. Select any other element of the I beam; note that the results given are relative between the top line and the selected element. You could select each of the other elements in turn and they will always be analyzed relative to the top line. When you no longer require the analysis to be relative to the top line, click on the YES button and you can start the analysis relative to two new elements.

Using the two previous analysis options, experiment by making various selections from the I beam drawing. You could also use the other two exercises created in the previous chapter or draw additional elements and analyze them until you become familiar with the function. You can also experiment by making the same ANALYSIS selections when in 2D SPACE mode as in 3D SPACE mode. Refer to Chapter 9 if you need a reminder on how to switch between 3D and 2D SPACE modes.

NOTE: *The INERTIA option from ANALYSIS will be covered after you have created a few solids in the next chapter.*

Summary

This chapter focused on using selected ANALYSIS function tools to obtain information about 3D geometry.

11 Creating Solids

The objective of this chapter is to provide an introduction to solid modeling. Required background information to CATIA solids is summarized prior to discussion of the SOLIDM and SOLIDE functions and the exercises.

Background on Solids

Solid Representations

In CATIA solids have two representations. First, *B-REP* (boundary representation) is the visual graphical representation, that is, what the user sees on the screen. Next is *CSG* (constructive solid geometry), the representation of the history of the solid, or the steps used to create it. The CSG is available in the form of a list or a CSG tree diagram. Both representations are always available. More detailed information on representations is presented later in the chapter.

Mock-up and Exact Solids

CATIA solids come in two forms: mock-up and exact. A mock-up solid, produced using the SOLIDM function, has an approximate B-REP excepting solids comprised solely of planar faces (e.g., a cube). Mock-up solids consist of adjacent planar facets whose contours are limited by straight edges.

For example, a solid cylinder produced using SOLIDM will have a polygonal contour. The accuracy of the contour compared to the theoretical circle is called the "discretization." The two principal approximation parameters are described below.

- SAG. Amount that a facet deviates from the theoretical curve.
- LIN-APP (line approximation). Number of facets per quadrant (one quarter of a full rotation).

Primitive mock-up solids, such as prisms, cylinders, cubes, and so on require less model space than the equivalent exact solids. They can be useful in the early stages of assembly design to establish overall space and access requirements, basic shape, mass, and cost of component parts before proceeding to full detail design.

Primitive mock-up solids are quick and easy to produce with the use of the same techniques as described below for exact solids. They can also be upgraded later to exact solids if necessary via the MODIFY | SOL TYPE option in the SOLIDE function.

An exact solid produced using the SOLIDE function has an exact B-REP. In brief, such solid does not have facets. The graphical visualization of an exact solid may appear faceted under certain visualization discretization settings in the GRAPHIC function.

The basic method of producing simple and primitive solids is the same in SOLIDM and SOLIDE, but the tools for managing and modifying an exact solid are far superior to those for mock-up solids. In addition, automatic extraction of drawings from solid models (covered in a later chapter) is far more effective from exact solids than from approximate solids. For these reasons, the rest of this book will address exact solids only, although a description of options available in the SOLIDM function is provided below.

SOLIDM

The SOLIDM function, available only in SPACE mode, is used to create, analyze, modify, and display mock-up (approximate) solids. Options are described in the next table.

Option(s)	Description
CREATE | CANONIC | CUBOID or CYLINDER or PRISM or REVOLUTN or SPHERE or CONE or TORUS or PIPE or PYRAMID	Create canonical primitives (defined only by geometric values and/or a single parameter construction element, called a "contour").
CREATE | COMPLEX | VOLUME	Create complex primitives through volume transformations.
CREATE | COMPLEX | OFFSET	Create complex primitives by offsetting surfaces, polyhedral surfaces, or non-planar faces.
CREATE | COMPLEX | PROJECT	Create complex primitives by projecting surfaces, polyhedral surfaces, or non-planar faces.
CREATE | COMPLEX | CLOSE	Create complex primitives by closing surfaces, polyhedral surfaces, or non-planar faces.
CREATE | SURFACE	Create a polyhedral surface.
CREATE | MACRO	Create macroprimitives.
OPERATION | UNION or INTERSEC or SUBTRACT or SPLIT or ASSEMBLE or SORT OUT	Perform operations.
ANALYZE | SELF | INERTIA	Analyze mechanical data of a solid.
ANALYZE | SELF | NUMERIC	Analyze modeling data of a solid.
ANALYZE | SELF | PARM	Analyze and create geometric parameters.
ANALYZE | POSITN | RELATIVE	Analyze relative position of two solids.
ANALYZE | POSITN | INTERFER	Analyze relative position of an unspecified number of solids.
ANALYZE | POSITN | INTERFER | COMPUTE	Compute interferences existing in a set of solids.
ANALYZE | POSITN | INTERFER | RELATN	Analyze interferences existing in a set of solids.
ANALYZE | POSITN | INTERFER | ELEMENT	Analyze interferences between a solid and a set of solids.
ANALYZE | POSITN | ALIGNMNT	Analyze tenon and mortise alignment.
MODIFY | GEOMETRY	Modify the geometry of a solid.
MODIFY | GEOMETRY | MOVE | TRANSLATE or ROTATE or SYMMETRY or UNSPEC	Apply a geometric transformation to a branch of a solid.
MODIFY | GEOMETRY | PARM	Modify the parameters of a primitive.
MODIFY | GEOMETRY | CONTOUR	Modify a primitive with a contour.
MODIFY | OPERATN | INSERT or DELETE | UNION or INTERSEC or SUBTRACT	Modify the history of a solid.
MODIFY | OPERATN | REPLACE or DUPLICAT | TRANSLAT or ROTATE or UNSPEC	Replace or duplicate a branch in the history of a solid.
MODIFY | VISUALTN | ELEMENT or EDGE	Modify the special display of solids.

Chapter 11: Creating Solids

Option(s)	Description
MODIFY I DRESS UP I RENAME I PRIM or FSUR or EDGE	Modify primitive attributes.
MODIFY I DRESS UP I COLOR	Modify the color of a primitive.
EXTRACT I CUT or PROJECT I EXACT or APPROXIM	Produce geometry by cutting or projecting solids.
EXTRACT I PRIMVOL	Extract subelements related to primitives forming solids.
VISU STD	Define the display standard.
UPDATE	Update a solid (i.e., bring the B-REP up to date with the latest CSG).
RESTORE	Restore a solid (i.e., return it to the state of the last update).

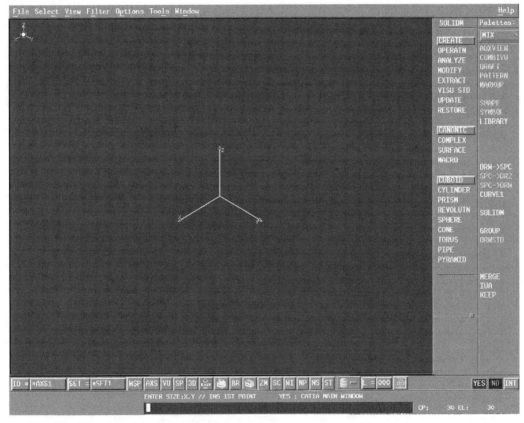

SOLIDM function menu.

SOLIDE

The SOLIDE function, only available in SPACE mode is used to create, analyze, modify and display exact solids.

Option(s)	Description
CREATE | CANONIC | PRISM or CYLINDER or REVOLUTN or SWEEP or CUBOID or SPHERE or CONE or TORUS or PIPE or PYRAMID	Create canonical primitives (defined only by geometric values and/or a single parameter construction element known as a "contour").
CREATE | COMPLEX | OFFSET	Create complex primitives by offsetting surfaces or faces.
CREATE | COMPLEX | PROJECT	Create complex primitives by projecting surfaces.
CREATE | COMPLEX | CLOSE	Create complex primitives by closing surfaces.
CREATE | COMPLEX | VOLUME	Create complex primitives by transforming volumes.
CREATE | FEATURE | CATALOG or BRANCH	Create primitives by using features.
CREATE | MACRO	Create solids with macroprimitives (solids made from dittos).
OPERATN | UNION | INTERSEC | SUBTRACT | THICK or DRAFT or FILLET or CHAMFER or SHELL or SPLIT or SEWING or SORT OUT or INACTIVE	Perform operations.
ANALYZE | SELF | INERTIA	Analyze mechanical data of a solid.
ANALYZE | SELF | NUMERIC	Analyze modeling data of a solid.
ANALYZE | SELF | PARM	Analyze and create geometric parameters.
ANALYZE | POSITN | RELATIVE	Analyze the relative position of two solids.
ANALYZE | POSITN | INTERFER	Analyze the position of an unspecified number of solids.
ANALYZE | POSITN | INTERFER | COMPUTE	Compute the interferences existing in a set of solids.
ANALYZE | POSITN |INTERFER | ELEMENT	Analyze interferences between a solid and a set of solids.
ANALYZE | POSITN | ALIGNMNT	Analyze tenon and mortise alignment.
MODIFY | GEOMETRY	Modify the geometry of a solid.
MODIFY | GEOMETRY | MOVE | TRANSLATE or ROTATE or SYMMETRY or UNSPEC	Apply a geometric transformation to a branch of a solid.
MODIFY | GEOMETRY | PARM	Modify the parameters of a primitive.
MODIFY | GEOMETRY | CONTOUR	Modify a primitive with a contour.
MODIFY | OPERATN | INSERT or DELETE or UNION or INTERSEC or SUBTRACT	Modify the history of a solid.

Option(s)	Description
MODIFY | OPERATN | REPLACE or DUPLICAT | TRANSLAT or ROTATE or UNSPEC or EDGE	Replace or duplicate a branch in the history of a solid.
MODIFY | DRESS UP | RENAME | PRIM or FSUR or EDGE	Modify primitive attributes.
MODIFY | DRESS UP | COLOR	Modify the color of a primitive.
EXTRACT | CUT or PROJECT | EXACT or APPROXIM	Produce geometry by cutting or projecting solids.
EXTRACT | PRIMVOL	Extract subelements related to primitives forming solids.
VISU STD	Define display standard.
UPDATE	Update a solid (i.e., bring the B-REP up to date with the latest CSG).
RESTORE	Restore a solid (i.e., return it to state prior to last update).

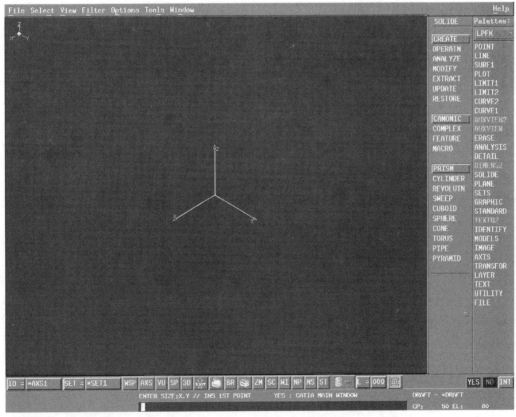

SOLIDE function menu.

NOTE: *As you are by now aware, there are many ways of achieving an end result in CATIA. Nowhere is this more true than with solid modeling. As you gain experience you will develop your own techniques. This chapter provides an introduction to techniques upon which you can build and experiment.*

Exercise 1: Creating a Complex Solid Using Primitives

This exercise is designed to provide insight into some of CATIA's solid modeling methodology before using the 3D geometry of the I beam and other beam trolley components. The exercise covers creating primitive exact solids, and performing operations and modifications. Analysis of solids is covered in Chapter 12.

The next illustration shows the end result of the solid to be created in this exercise. Step-by-step instructions are provided for every part of the solid, beginning with construction of the base.

Target solid.

Solid Base

1. Select SOLIDE > CREATE | CANONIC | CUBOID. In the Manage panel, select the Create a new feature option.

> **NOTE:** *As you pass the cursor over the icons, respective functions are displayed in messages.*

2. The REF PT: MSELW PT // DIR: SEL LN/PLN prompts display. The prompts can be translated as follows: choose a reference point or multi-select several points if several solids are to be produced, and then select a line or a plane to determine the direction of the cuboid.

3. Key in ,, and press <Enter> to use the origin as the reference point for the cuboid. The SEL ARROW // SEL PT // SEL DIM KEY (LX,LY,LZ) // YES: CREATE prompts display. These prompts mean select an arrow dimension to modify; select a point; select a dimension; key in values for LX, LY, and LZ; and click on the YES button to create.

4. The base will be 150mm x 75mm x 25mm thick. Select the LZ dimension in the work area, key in *25*, and press <Enter>. Select the LX dimension in the work area or in the CUBOID panel, key in *150*, and press <Enter>. Select the LY dimension, key in *75*, and press <Enter>. Finally, click on the YES button to create the cuboid.

Wireframe cuboid.

Exercise 1: Creating a Complex Solid Using Primitives

Cuboid dimension panel.

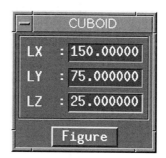

SOLIDE Manage panel.

Conical Solid

1. Select SOLIDE > CREATE | CANONIC | CONE. Prompts display as mentioned previously.

2. Key in *75,37.5,25* and press <Enter>. This is the coordinate of the anchor point for the conical solid on the center of the upper surface of the base solid.

3. A prompt displays requesting the direction. Select the Z axis and then enter the following values in the CONE panel: R1 = *20* (base radius), or if the D1 symbol is showing, enter a diameter or click on D1 in the work area to switch to R1; R2 = *10* (top radius); H2 = *50* (height); and H1 = *0*.

4. Click on the YES button to create the cone.

Exercise 1: Creating a Complex Solid Using Primitives

Sphere

1. Select SOLIDE > CREATE | CANONIC | SPHERE. For the position of the sphere center, key in *75,37.5,145* and press <Enter>.

2. Enter a value of *25* for the radius, and then click on the YES button to create the sphere.

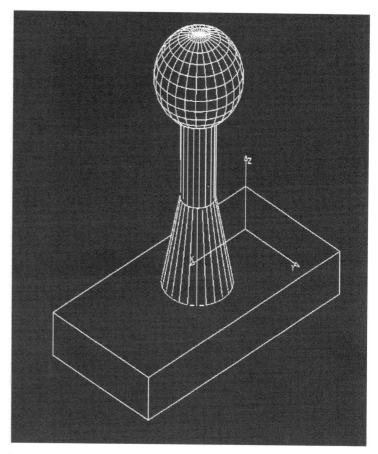

Base, cone, cylinder, and sphere.

Drafting Base Solid Sides

Drafting in this context refers to the creation of angled faces to facilitate the removal of the component from a mold (as in a casting or molding). In this context drafting is not related to the 2D DRAFT function in any way. (See Chapter 6 on 2D DRAFT function.)

1. Select SOLIDE > OPERATN | DRAFT. A central arrow indicating direction of pull displays.

   ◆ *NOTE*: *During the DRAFT process, a plane is automatically generated. Use ERASE > NOSHOW to hide it from view.*

2. Select one side of the base. An arrow appears. To change the draft angle to 5 degrees, key in *5* and press <Enter>. Repeat the entry for the remaining three sides.

3. Click on the YES button to perform the draft operation.

Joining Separate Primitives to Form Single Solid

1. Select the base solid.

2. Key in *SOL (multi-select, meaning all solids) and press <Enter>, or select each of the solids in turn. The NUMBER OF BODIES = 1 message displays indicating that the four solid primitives have been joined.

Fillet Radius Around Top Edge of Base

1. Select SOLIDE > OPERATN | FILLET | EDGE. For the fillet radius, key in *5* and press <Enter>.

2. Select the top surface of the base.

   ◆ **NOTE:** *Four display modes are available by clicking on the Display mode button on the fixed menu bar. (To locate the button, move the cursor over the buttons to view messages describing the purpose of each button.) The display modes follow: NHR—no hidden line removal (all solids are transparent); HLR—hidden lines removed without dynamic buffer regeneration; HLR—hidden lines removed with dynamic buffer regeneration; and Shaded/Rendered image with dynamic buffer regeneration.*

3. To select an FSUR (the face of an exact solid) in NHR mode, select one of the dimmed lines lying on the selected surface. To select a face in dynamic HLR or shaded mode, select any point in the area of the selected surface.

Exercise 1: Creating a Complex Solid Using Primitives 233

4. The following prompts will display: STD: RADIUS 3.000 SEL REDG / VERTEX / RSUR // SEL RAD // KEY RAD // YES: COMPUTE. Select the four corner edges of the base. Click on the YES button to perform the fillet operation.

➥ **NOTE:** *Because the face of the base solid was selected, a fillet radius is created where the surface meets the cone.*

Complete solid with fillets and draft.

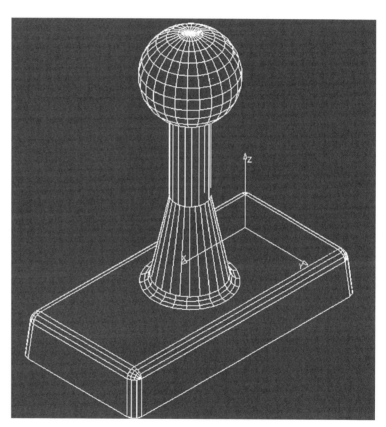

Performing Local Transformations of the 3D Model

The following steps describe how to twist, turn, and drag the 3D model for purposes of allowing you to view it from any direction.

1. Hold down mouse button 3 and press button 1 (or press <F4>). The ensuing window provides access to various tools for changing the orientation of the 3D image on the screen. To switch the window off, simply repeat the same mouse action or select the off switch in the window.

2. If the word "locked" appears in the bottom half of the window, confirm that you have not activated the window by clicking on the circle on the axis icon at the top left corner of the screen. It should be highlighted. If the circle is highlighted and the locked message continues to display, select STD from the window and verify that the flat window switch is set to off.

3. Select 3D in the window and then select UNSPEC ROTATION. Choose the bottom front edge of the base solid; a red dashed line appears. Assuming that your setup includes dials, the solid will rotate about the selected line as you turn dial 4.

4. When you press and hold mouse button 2, the virtual spaceball will appear. The ball's center lies on the midpoint of the selected rotation axis. A cross on the ball surface represents where your hand is in contact with the ball. While holding down button 2, move the mouse backwards and forwards and from side to side and the ball will appear to roll rotating the model. Experiment with the spaceball; you will be surprised how quickly you get the hang of it.

Exercise 1: Creating a Complex Solid Using Primitives

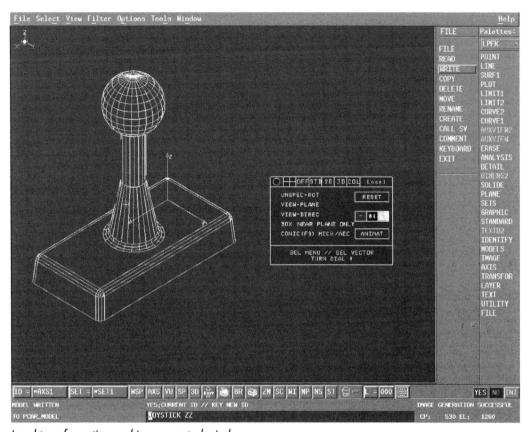

Local transformation and image control window.

Create Shell

1. Using the method described above, rotate the model so that you can see the underside of the base.

2. Select SOLIDE > OPERATN | SHELL, and then select the solid. The following prompts display: REMOVED FACES: SEL FSUR THICK: KEY OFF(,OFF2) // YES: CURR.

3. Select the bottom face of the base solid. This face will be removed to produce an open shell. For a shell thickness of 2mm (.078") inside the external shape, key in *-2* and press <Enter>.

4. Click on the YES button to compute the shell operation. Using the virtual spaceball, manipulate the model in one of the dynamic display modes to view the shell from various angles.

5. Save the model for later practice sessions.

Shell solid.

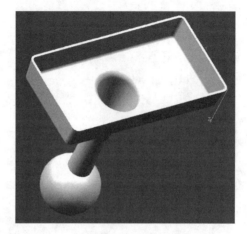

Cutting Shell Solid in Half

In order to view the inside of the shell more clearly and to clarify subsequent operations, the solid will be cut in half. Take the following steps.

1. Select PLANE > PARALLEL. To produce a plane on the center line of the base, key in *xz* and press <Enter>, and then key in *37.5* and press <Enter>.

2. Select SOLIDE > OPERATN | SPLIT. In the Mode window select One Side and Trim.

3. Select the plane. The ensuing arrow should point toward the axis. If it does not, select the arrow to invert it. The following prompts display: MSELW/SOL/DIT/VOL INVERT:SEL ARROW.

4. Select the solid. The solid will be cut in half.

➥ **NOTE:** *To manipulate the model with the virtual spaceball without accessing the window, press and hold mouse buttons 2 and 3 simultaneously.*

Modifying Solid Parameters

Steps to change shell thickness follow.

1. Select MODIFY > GEOMETRY | PARM. In the Manage window select the Display part editor button to access the Part editor window and the CSG tree. Select the SH (shell) icon from the CSG tree, or select the shell surface in the work area. The following prompts display: OFFSETS = -2.000,0 THICKNESS VALUES: KEY OFF1 (,OFF2).

2. To change the shell thickness to 4mm, key in *-4* and press <Enter>. The background of the CSG window turns green, which means that the solid B-REP must be updated.

3. Select the Update the Current solid button in the Part editor window. Shell thickness will change to 4mm.

The sphere radius will be modified in the following steps.

1. Select MODIFY | GEOMETRY | PARM and access the Part editor.

2. Select the SP (sphere) icon from the CSG tree, or select the sphere in the work area. Click on the 25.000 dimension in the Sphere window, and then key in *40* and press <Enter>. The graphic of the sphere will change and the following prompt will display: SEL SYMB SEL CENTER // YES: MODIFY.

3. Click on the YES button to execute the modification.

4. Click on the Update the current solid button, or select UPDATE from the menu and click on the YES button. The radius of the sphere changes to 40mm.

Changing the base depth is the final modification. Take the following steps.

1. Select MODIFY | GEOMETRY | PARM and access the Part editor.

2. Select the CU icon (cuboid) from the CSG tree.

3. Double-click on the 25 dimension in the CUBOID panel, and then key in *50* and press <Enter>.

4. Click on the YES button to execute the modification.

5. Click on the Update the current solid button. Base thickness has changed to 50mm and the height of the cone has been reduced to accommodate the change.

➥ **NOTE:** *All operations performed above can be deleted by selecting SOLIDE > MODIFY | OPERATN | DELETE, and then selecting branches from the CSG tree or selecting the relevant primitive in the work area.*

Modified solid.

The series of operations in this exercise provide insight into several methods for building and modifying solids. The possibilities are infinite. Experiment with the solid created here by making additional changes, adding solids, and creating new solids until you become familiar with the new tools.

You can also build and modify solids using SOLIDE > CREATE | FEATURE where branches of the CSG tree can be reused, or predefined catalog features can be read from a library. Check with your system administrator to determine whether such facility is available.

Exercise 2: Creating a Solid I Beam

To work with this exercise, read the model of the I beam 3D geometry saved in Chapter 9. The solid I beam will be created as an exact solid prism 600mm long.

Exercise 2: Creating a Solid I Beam

1. Select SOLIDE > CREATE | CANONIC | PRISM. The following prompts display: CONTOUR: MSELW LN/CONIC/CRV/CCV CONTOUR: SEL PT/ FAC. Verify that the No associative limits option is selected in the PRISM window, and that the Create a new feature option is selected in the Manage window.

2. Select one element of the I beam contour, and then click on the AUTO-SEARCH button in the Contour window. The complete contour will be highlighted.

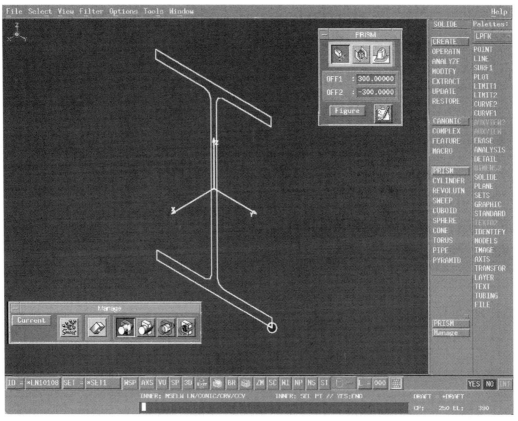

Creating a prism.

3. Click on the YES button to end the contour selection. The following prompts will be displayed: SEL ARW // SEL ELEM // SEL DIM KEY(OFF1,OFF2) // YES: CREATE.

4. Key in *300,-300* (the limits of the prism) and press <Enter>. The prism is created and the message CURRENT SOLID CHANGED displays.

At this stage, transferring the solid to a different layer is recommended in order for the solid sans geometry to be seen. Take the following steps.

1. Select LAYER > LAYER | TRANSFER. The following prompts display: SEL > ELEM // SEL LAYER KEY LAYER // YES: CURRENT.

2. Key in *10* (to choose layer 10) and press <Enter>, and then select the solid. The solid will change to no-pick, which means that the solid has been transferred to layer 10.

3. Click on the L= (layer) button on the fixed menu bar. The following prompts display: YES: FORWARD // NO BWD KEY NUM // SEL: ELEM. Key in *10* and press <Enter>. Layer 10 becomes the current layer button on the fixed menu bar.

4. Apply a filter so that only layer 10 is visible. Select LAYER > FILTER | APPLY. Select LAYCUR (current layer filter) from the Filters window. Note that the I beam geometry disappears; only layer 10 is visible. *This procedure should be followed at the end of Exercises 3 through 5.*

5. Save the model as *BEAM TROLLEY SOLID I BEAM*.

Exercise 3: Creating the Solid Side Plate

Read the model of the I beam 3D geometry saved in Chapter 9. The side plate will be created as an exact solid prism 10mm (.394") thick, which will then be modified to add holes and the edge chamfer.

1. Select SOLIDE > CREATE | CANONIC | PRISM. Settings in the Prism and Manage windows should match Exercise 1 settings.

2. Select part of the side plate contour, and now choose another element of the contour not connected to the one already selected. The complete contour is highlighted. (This method of selecting a contour is an alternative to using the AutoSearch button.)

Exercise 3: Creating the Solid Side Plate

3. Click on the YES button to end the contour selection. Red arrows indicating the direction of the extrusion of the prism appear. To create a solid that is 10mm (.394") thick, either key in *,10* and press <Enter>, or modify the values in the PRISM window as follows: OFF1 = 0, and OFF2 = 10.

4. Transfer the solid to layer 10.

The next stage in the creation of the side plate is to create the four holes. Take the following steps.

1. Select SOLIDE > CREATE | CANONIC | CYLINDER. In the CYLINDER window, click on the Until from to button which will enable the faces of the created prism (i.e., the current solid) to be used as the limits of the cylinder.

2. Click on the Remove from current solid (subtract) button in the Manage window. This selection means that as the cylinder is created it will automatically be subtracted from the side plate solid, thus creating the first hole.

3. Select one of the four circles representing the geometry of a hole.

→ **NOTE:** *Verify that all lines are visible by selecting the no hidden line removal mode using the display mode button on the fixed menu bar. You will then be able to see the FSUR (functional surfaces) of the solid represented by dimmed lines.*

4. Select the FSUR at the rear of the solid. The FSUR will be highlighted and LIM1 displayed at its center.

5. Select the FSUR of the front face of the side plate; it too will be highlighted with LIM2 displayed.

6. Click on the YES button to create the hole.

The next stage is to create the other three holes. The subtraction of the first hole from the side plate solid is called an "operation." Click on the Display part editor button in the Manage window to display the CSG tree. Note that the solid now consists of a PRISM branch and a CYLN branch joined at the subtraction (-) icon, which signifies that the subtraction operation was performed.

To create the three remaining holes you are going to duplicate the subtraction operation at the other hole positions. Take the following steps.

1. Select SOLIDE > MODIFY | OPERATN | DUPLICAT | TRANSLAT. The SEL BRANCH prompt displays.

2. The hole can now be selected by either selecting its FSUR in the work area, or by selecting the CYLN branch in the CSG tree. Select the hole using one of these methods, and then select the center point of the highlighted hole. Choose the center point of one of the other holes; a red arrow representing the defined translation displays.

3. Click on the YES button to apply the translation. Note that the background of the CSG tree window turns green, meaning that the solid's BREP (visual appearance) must be updated. The update will occur when all holes have been created.

4. To produce the next hole reselect the highlighted FSUR of the new hole, select its center point, and then choose the center point of one of the other circles. Click on the YES button to apply.

5. Repeat the above procedure to create the fourth hole and then click on the Update the current solid button in the Part Editor window. The four holes have now been created. Note the four CYLN branches in the CSG tree.

Exercise 3: Creating the Solid Side Plate

Side plate solid with four holes.

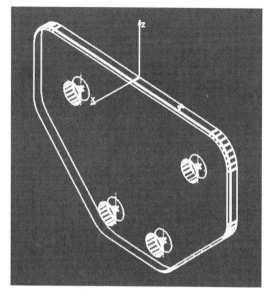

The last step in the creation of the finished side plate is to generate a 2 x 2mm chamfer around one of its edges.

1. Select SOLIDE > OPERATE | CHAMFER. In the Chamfer mode window, verify that the propagation switch is set to Auto, and the Parm mode switch to L1,L2.

2. Double-click in the L1 entry area, key in *2*, and press <Enter>. Repeat for L2.

3. Select the nearest edge of the side plate. L1 and L2 are displayed in the work area, and the edge to be chamfered is highlighted by a red dashed line with small circles at the change of direction.

4. Click on the YES button to create the chamfer.

5. Transfer the solid to layer 10.

6. Save the side plate as *BEAM TROLLEY SOLID SIDE PLATE*.

Exercise 4: Creating a Solid Wheel

Read the 3D geometry model of the wheel created in Chapter 9. The geometry will be used to create an exact revolution solid, and the solid will be modified to add webs and fillets.

1. Select SOLIDE > CREATE | CANONIC | REVOLUTN. The following prompts display: SEL REF PT // AXIS: SEL LN/CIR/PLN. Select the center line for the axis of rotation.

2. Select one element of the solid line contour, and then click on the Auto-Search button in the Contour window. Verify that A1 is set to *0* and A2 is set to *360*. Click on the YES button to create the solid.

First stage of the solid wheel.

The steps to create the webs follow.

1. Select SOLIDE > CREATE | PRISM. Settings in the Prism and Manage windows should match Exercise 1 settings. Select one element of the dotted geometry of the web, and click on the AutoSearch button. Click on the YES button to end the contour.

Exercise 4: Creating a Solid Wheel

2. Key in *3,-3* and press <Enter> to create the solid web. Note that the web geometry overlaps the wheel geometry by 1mm nearest the center line. The overlap accounts for the curvature of the hub and ensures full width contact of the web when the union operation is executed.

3. Select SOLIDE > OPERATE | UNION. Verify that the Trim all mode is selected. Select the wheel and then select the web. Click on the Display the part editor button in the Manage window. (Note that the solid has two branches.)

4. To produce the remaining two webs, select SOLIDE > MODIFY OPERATN | DUPLICAT | ROTATE. From the CSG tree select the PRSM branch, and select the Y axis for the rotation axis. In the Rotation window, key in *120* and press <Enter>.

✗ **WARNING:** *If by mistake you enter 120 in the normal information entry area, you will get 120 webs at the angle currently showing in the Rotation window. Be careful.*

5. Click on the YES button to apply, and then click on the YES button again to iterate the operation.

6. Click on the Update the current solid button in the Part editor window to update the B-REP of the solid.

The next stage is to add fillets and chamfers. Take the following steps.]

1. Select SOLIDE > OPERATN | FILLET. Select each edge where a fillet is required, and key in *3* and press <Enter> for a 3mm radius. Click on the YES button to execute the operation.

2. To add chamfers, select SOLIDE > OPERATE | CHAMFER. Change the values of L1 and L2 to 1mm, and then select all the outer edges of the wheel. Click on the YES button to execute the operation.

3. Click on the Update button to update the B-REP and reveal the completed solid wheel. Transfer the solid to layer 10.

4. Save the model as *BEAM TROLLEY SOLID WHEEL*.

Completed solid wheel.

Exercise 5: Creating a Solid Nut

This exercise demonstrates the method of using two solids to create a single intersection solid. First, however, nut geometry is required. Take the following steps.

1. Create a new model with FILE > CREATE | YZ. Click on the 3D button on the fixed menu and then select the X axis. You are now ready to draw in the YZ plane.

2. Select CURVE2 > CIRCLE | DIAMETER. Key in , and press <Enter> to draw a circle on the origin. Key in *34.65* and press <Enter> to draw a 34.65 diameter circle which is the across points dimension of an M20 nut.

3. Repeat step 2 to draw a 20 diameter circle.

4. Select POINT > SPACES | PTS. Select the circle, and then key in *5* and press <Enter>. Click on the YES button to use the end points. Now select POINT > LIMITS | PTS, and select the circle again to create a point on the circle limits and the center point.

5. Select LINE > PT_PT | SEGMENT | STANDARD. Select one of the points on the circle. Note that rubberbanding is activated. Select an adjacent point and then click on the YES button to start Chain mode.

Exercise 5: Creating a Solid Nut

6. Work around the circle selecting the points in turn until the hexagon is complete. After selecting the final point, click on the YES button to end Chain mode.

7. Click on the 2D button on the fixed menu, and then click on the YES button to return to 3D mode.

8. Select POINT > PROJ_INT | SINGLE | LIM OFF. Select the Z axis and then the top side of the hexagon to create a point. Click on the 3D button on the fixed menu and then select the Y axis. You are now ready to draw in the XZ plane.

9. Select LINE > SEGMENT | ONE LIM. Select the point that you just created, key in *-30* and press <Enter>, and then key in *-25* and press <Enter>. A line 25mm long at 30 degrees from the X axis is created. The geometry for the nut is now complete. Click on the 2D button on the fixed menu, and then click on the YES button to return to 3D mode.

Nut geometry.

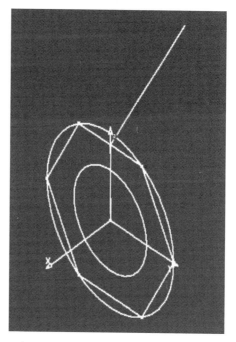

Now that the geometry is complete, you can proceed to produce the solid nut. Take the following steps.

1. Select SOLIDE > CANONIC | PRISM. Select one of the six sides of the hexagon, and then click on the AutoSearch button. The following prompts display: INNER: MSELW LN/CONIC/CRV/CCV INNER: SEL PT // YES: END.

2. Select the diameter 20 circle and then click on the YES button to end the inner contour. Key in *,-16* and press <Enter> to produce a hexagon solid with a 20mm hole through it.

3. Select SOLIDE > CREATE | CANONIC REVOLUTN and then select the X axis for the rotation axis. Select the 30 degree line and click on the YES button. Verify that A1 = 0 and A2 = 360, and then click on the YES button to create the revolution solid.

4. Change to shaded display mode. Select SOLIDE > OPERATE | INTERSEC. Select the hexagon solid, and then select the revolution solid. The finished nut appears. The result is a solid comprised only of the volume common to both the previously selected solids (their intersection).

5. Transfer the solid to layer 10.

6. Save the model as *BEAM TROLLEY SOLID NUT*.

Complete solid nut.

The remaining solids required for the complete beam trolley are the wheel spindle, stud, bearings, bushes, and washers. Using the techniques discussed in Chapters 9 through 11, create these solids to gain experience. If you encounter problems, the starting geometry and finished solids models can be read from the companion CD. This chapter has introduced various solid modeling techniques and tools. Experiment with them and others in the menu to discover dif-

Summary

ferent methods of producing solids. Remember that although a single correct method does not exist, you should take into account how the model will be used in the future.

✓ **TIP:** *The spindle, studs, bearings, bushes, and washer will be revolution solids, while the other parts will be cylinders or prisms.*

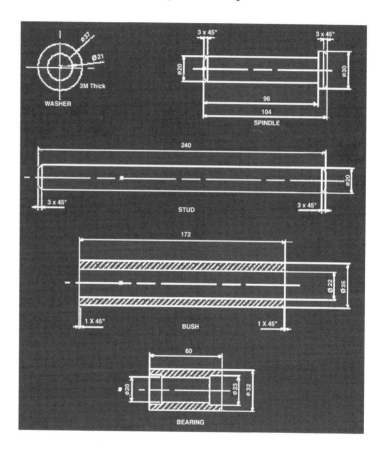

Additional beam trolley solids geometry.

Summary

This chapter focused on creating primitive solids (cuboids, cones, cylinders, spheres, and prisms), joining solids to make more complex solids, and performing operations on solids (draft, fillet, shell, union, subtract, intersect, and split).

12 Analyzing Solids

Analysis of solids is essential. For example, you may wish to check the physical dimensions of one or more solids and the relative distance between two solids, as well as whether interferences or clashes within a group of solids occur. You may also need to know the mass, surface area, center of gravity, or inertia of one or more solids. CATIA provides the following two analytical methods:

- ANALYSIS function (introduced in Chapter 10). This method will be used in the first exercise to obtain information about the I beam solid, and again in the final exercise to produce the combined analysis of several solids.

- ANALYSIS option in SOLIDE function (introduced in Chapter 11). This method is introduced in Exercises 2 and 3 to analyze parameters of selected solids that you have produced, and to perform relative position analysis and clash detection for a group of solids.

Exercise 1: Analyzing Solids Using the ANALYSIS Function

Read the *BEAM TROLLEY SOLID I BEAM* model saved in Chapter 11. Select No Hidden Line Removal (top option) by clicking on the Set current display mode button on the fixed menu.

Chapter 12: Analyzing Solids

➥ **NOTE:** *Verify that you are in SPACE mode. If SP is not displayed, click on the DR button on the fixed menu to switch to SP.*

1. Select ANALYSIS > NUMERIC | COMPUTE | RELATIVE.

2. Select an FSUR type element (i.e., one of the dimmed lines representing a planar face of the solid). A highlighted PLANE symbol is displayed in the work area and the equation of this plane is displayed in the alphanumeric window.

3. Select an edge line of the solid. The element is highlighted in the work area and the following characteristics are displayed in the alphanumeric window: starting point, end point, vector, supplement angles between the selected line and the three main planes, and length.

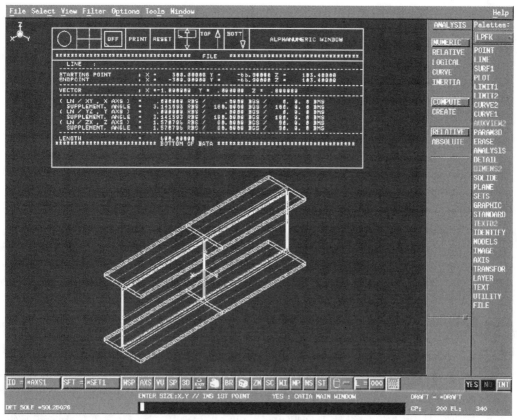

Typical analysis window.

Exercise 2: Analyzing Solids Using the SOLIDE Function's ANALYSIS Option

4. Select one of the FSUR type elements of the filleted corner of the I beam solid. The INVALID FOR NUMERICAL ANALYSIS message is displayed. If an FSUR element is selected (other than those on a curved surface), the associated plane will be selected as before; if a second FSUR is selected, the relationship between the two planes will be analyzed; and if an edge line is selected, the analysis of the highlighted line will be displayed.

5. Choose ANALYSIS > LOGICAL | STANDARD, and then select any element. The alphanumeric window informs you that a solid with a single body is detected and that no application is using the element.

6. Select ANALYSIS > LOGICAL | CHILDREN, and then select any element. A list of elements with respective IDs is displayed.

7. Choose ANALYSIS > LOGICAL | PARENTS, and then select any element. A list of the elements used to create the solid that are logically linked to the selected element is displayed.

8. Select ANALYSIS > LOGICAL | FAMILY, and then select any element. A list of elements identified by type that belong to the family of selected solids is displayed.

9. Choose ANALYSIS > LOGICAL | CURVE. Only CCV, PIP, and NET type elements are allowed; therefore, this option is not suitable for solid analysis.

10. Select ANALYSIS > INERTIA. For inertial analysis, the INERTIA option in the SOLIDE function will be used. The ANALYSIS function will be reintroduced later in this chapter to combine the inertial analysis of a group of solids.

Exercise 2: Analyzing Solids Using the SOLIDE Function's ANALYSIS Option

1. Select SOLIDE > ANALYZE | SELF | INERTIA, and then choose the solid. The solid is highlighted, the Manage window is displayed, and an additional Analysis window is also displayed. In the latter window, you must enter a value for material density.

➡ **NOTE:** *If you are interested only in the position of the solid's center of gravity, the density can be left unchanged.*

2. For this example, and assuming that the CATIA session is set up for SI units, double-click in the Density entry area, key in *0.0078* (the density for steel in gms/mm^2) and press <Enter>. During the analysis, temporary elements will be created to indicate the position of the center of gravity and the axes of inertia. You can retain these elements by selecting Keep Elements in the Analysis window. Next, for purposes of this exercise, the Density Cube should not be selected. If the check mark is showing, click on it to cancel. Finally, click on the YES button to compute the analysis.

3. The alphanumeric window should be displayed automatically. If it does not display, hold down the <Alt> key and press the plus symbol (+) key, or hold down mouse button three and press on button one. In the window that displays, select STD and ALPHA WIND ON. The alphanumeric window displays the following characteristics: solid ID; volume; mass; wetted area; center of gravity; and first, second, and third axes of inertia; inertia I1, I2 and I3, and inertia moments X,Y, and Z.

4. Select SOLIDE > ANALYZE | SELF | NUMERIC, and then select any element. The alphanumeric window displays the following: solid ID; number of vertices, edges, faces, primitives, and bodies; and index and data size.

5. Select SOLID > ANALYZE | SELF PARM, and then select any element. The alphanumeric window displays direction and limit planes of the prism.

Exercise 2: Analyzing Solids Using the ANALYSIS Option

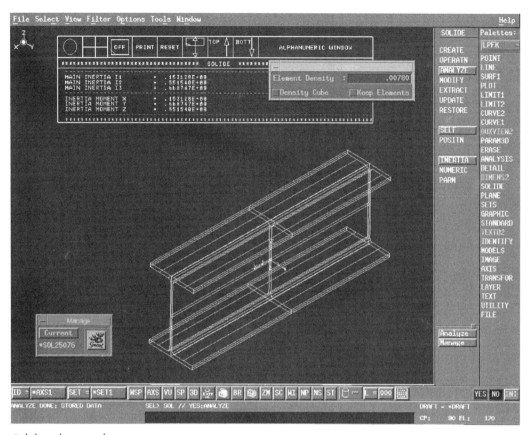

Solid analysis results.

Before proceeding, read the *BEAM TROLLEY SOLID WHEEL* model saved in Chapter 11. Because this model is a more complex solid, it will be more useful in demonstrating the parameter analysis. With the solid wheel on the screen, take the following steps.

1. Select SOLIDE > ANALYZE | SELF | PARM, and then select the main body of the wheel, either in the work area or from the CSG tree in the Part Editor window. The selected primitive is highlighted. The parameters are displayed in the work area and in the REVOLUTION window. The alphanumeric window displays the information shown in the next illustration.

Chapter 12: Analyzing Solids

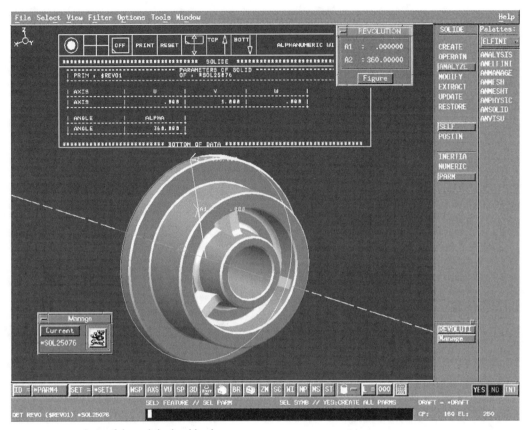

Parameter analysis of the solid wheel body.

2. Select one of the wheel webs. Parameters of the web primitive are displayed.

3. Select one of the fillet radii. The values of the fillet radii display in the alphanumeric window, as well as the PRIM WITH NO GEOMETRIC PARMS message. The fillets were created as an operation and do not contain geometry.

Exercise 3: Analyzing the Position of Several Solids

Although the TRANSFOR (transformation) function in SPACE mode will not be covered in detail until Chapter 13, the following steps will allow you to perform the necessary simple translations. Continue with the *BEAM TROLLEY SOLID WHEEL* model and proceed as follows.

1. Select TRANSFOR > CREATE | TRANSLAT. Key in *x* and press <Enter> to choose the x axis as the direction of the translation. Key in *150* and press <Enter>; a red arrow indicating the defined translation appears.

2. Choose TRANSFOR > APPLY | DUPLICAT | SAME, and then select the wheel. A duplicate solid will be created.

3. Select TRANSFOR > CREATE | TRANSLATE. Key in *z* for the translation direction. Key in *100* and press <Enter> to define the translation.

4. Select TRANSFOR > APPLY | DUPLICAT | SAME, and then select the original wheel solid. Another duplicate solid is produced which obviously intersects with the other.

5. To perform position analysis, select SOLIDE > ANALYZE | POSITN | RELATIVE. Select the original solid and then the solid you moved along the x axis by 150mm.

Chapter 12: Analyzing Solids

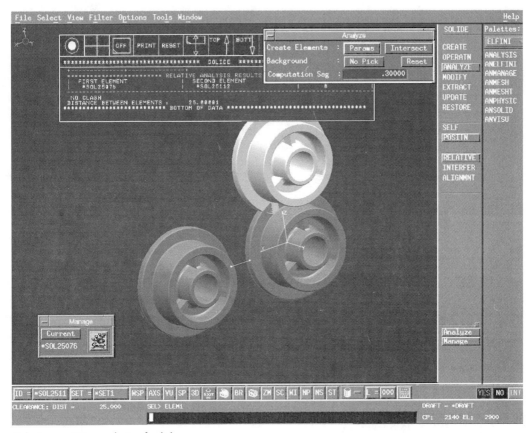

Relative position analysis of solids.

The clearance distance is displayed in the alphanumeric window as well as in the message area. A vector with endpoints is displayed indicating where the minimum clearance occurs. This vector can be retained by clicking on the Params button in the Analysis window. Descriptions of the other buttons in the Analysis window when used at this juncture appear below.

- Pressing the Intersect button will result in the following message: NO INTERFERENCE.

- When clicking on the No Pick button, everything except the solids selected for analysis will be placed in No Pick.

- Clicking on the Reset button reverses the previous step.

Exercise 3: Analyzing the Position of Several Solids

- Computational Sag represents the deviation of the B-REP of the exact solids from the theoretical shape for analysis purposes. You can enter a new value upon double-clicking in this area.

To perform interference analysis between the two solids that clash, take the following steps.

1. Select SOLIDE > ANALYZE | POSITN | INTERFER | COMPUTE. The MSELW > ELEM prompt asks you to multi-select the solids. Key in *sol and press <Enter>. Click on the YES button to end the selection. The 1 INTERFERENCES DETECTED message is displayed.

2. Choose SOLIDE > ANALYZE | POSITN INTERFER | RELATN. In the Interference List window, select *Solid1*. The two solids involved in the clash are highlighted. The details of the clash are displayed in the alphanumeric window and the message area.

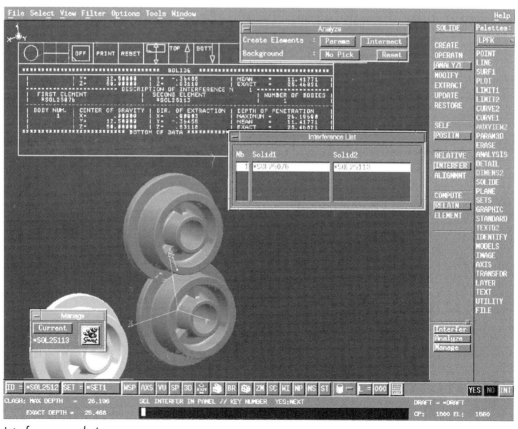

Interference analysis.

3. By clicking on the Intersect button, a solid is created which represents the volume of interference. This solid can be seen by placing the two solids being analyzed into No Show.

4. By selecting SOLIDE > ANALYZE | POSITN | INTERFER | ELEMENT, and then selecting one of the solids listed in the First Operand window, the selected solid is highlighted in the work area and the ID of the second solid is shown in the Second Operand window. The last option, SOLIDE > ANALYZE | POSITN | INTEFER | ALIGNMNT, is used to analyze the alignment of holes and pins, and other cylindrical alignments.

➥ **NOTE:** *If the message ERROR SEARCH FAILED is displayed at any point during the above procedures, select another function from the palette, re-select SOLIDE and the required option, and try again.*

Exercise 4: Combining Analysis Results

The following exercise demonstrates how to store inertial analysis results of several solids in order to combine them for the analysis of an assembly.

1. Using the three-wheel model, select ANALYSIS > INERTIA | COMPUTE | RELATIVE. Key in .078 for density and press <Enter>.

2. The inertial analysis of the selected solid is displayed in the alphanumeric window. Click on the YES button to store the results. Now select the next wheel solid, key in the density value again, and press <Enter>. Click on the YES button to store.

3. Repeat steps 1 and 2 for the third wheel.

4. Select ANALYSIS > INERTIA | COMBINE, and select each solid in turn. The inertial analysis results are combined, providing mass, C of G, and so forth for the group of solids.

Summary

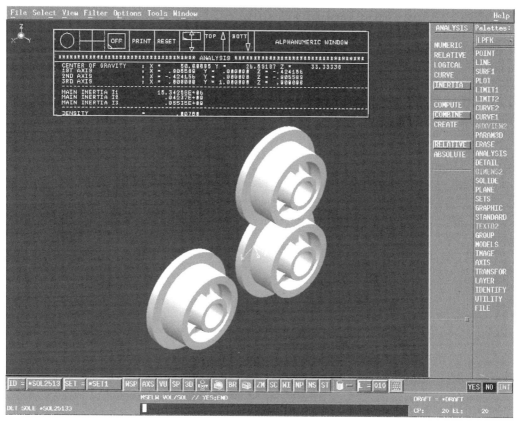

Combined inertial analysis of a group of solids.

Summary

This chapter focused on performing numerical and inertial analysis on a solid, relative position analysis of a group of solids, interference analysis between clashing solids, and combining analysis results.

13 Managing Solids

In the last two chapters several solids were created and analyzed, some of which are the component parts of the beam trolley assembly. The objective of this chapter is to manage these solids by moving them, merging some of them into a single model, creating duplicates of selected parts, and generally preparing the models for later creation of a multi-model environment.

In this chapter it will be necessary to re-read the beam trolley solids saved in Chapter 11. (You could also load all component part models from the Companion CDs.) The only new function introduced in this chapter is MERGE. Other functions employed in the exercises were covered Chapter 5, "Managing 2D DRAW Elements."

Options available in SPACE mode for each of the functions used in this chapter are described below.

TRANSFOR in SPACE Mode

Option	Description
CREATE | TRANSLATE	Create a translation.
CREATE | ROTATE	Create a rotation.
CREATE | SYMMETRY | PLANE or LINE or POINT | NORMAL or OBLIQUE	Create a symmetry.
CREATE | SCALING	Create a scaling.
CREATE | AFFINITY | PLANE or LINE or AXIS | NORMAL or OBLIQUE	Create an affinity (scaling in a single direction).

Chapter 13: Managing Solids

Option	Description
CREATE I PROJECT I NORMAL or OBLIQUE	Create a projection.
CREATE I MOVE	Create a move (combined translation and rotation).
CREATE I UNSPEC	Create an unspecified transformation by creating a three-axis transformation system.
CREATE I ANGLES	Create a move with three angles.
CREATE I EULER	Define a move by Euler angles.
APPLY I REPLACE or DUPLICATE I ELEMENT or SET or FAMILY I STANDARD or SAME	Apply a transformation.
STORE	Store a created transformation.
MANAGE I ANALYZE or ERASE or INVERT or COMBINE	Manage a transformation.

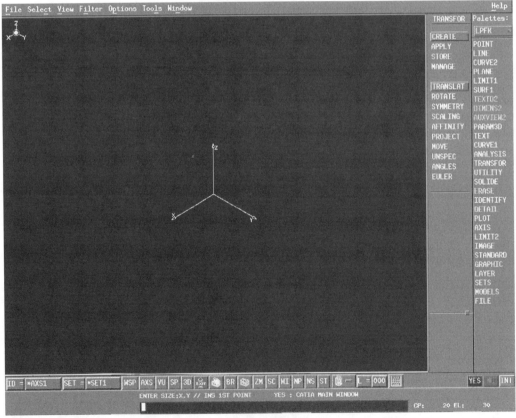

TRANSFOR (transformation) function menu in SPACE mode.

DETAIL in SPACE Mode

The DETAIL function in SPACE mode is used (as in DRAW mode) to place elements into an alternative workspace. The latter workspace can be used on the master workspace as a ditto or a copy of its elements. Options for performing transformations and analyses on dittos and to change between workspaces are described below.

Option	Description
DITTO | MODEL or LIBRARY | STANDARD or COMPACT	Create a ditto from existing detail in model or library.
COPY | MODEL or LIBRARY | STANDARD or COMPACT	Copy elements of an existing detail in the model or library in elemental form.
MODIFY | TRANSLATE or ROTATE or SCALE or SYMMETRY | REPLACE or DUPLICATE	Perform transformations on dittos.
EXPLODE | CURRENT or W.SPACE or MODEL	Transform a ditto into an elemental geometric copy.
CREATE	Create a new detail.
DELETE | UNUSED or USED | VISUALATN or DIRECT	Delete a detail.
MANAGE | ANALYZE | DETAIL or W.SPACE	Analyze a detail or a workspace.
MANAGE | UPDATE	Update a library detail with respect to the current library.
MANAGE | REPLACE | CURRENT or W.SPACE	Replace one ditto with another.
MANAGE | VERIFY | CURRENT or W.SPACE	Verify the nature of a ditto.
MANAGE | LAYER | CURRENT or W.SPACE	Change a standard ditto to a compact ditto or vice versa.
MANAGE | RENAME	Rename a detail.
MANAGE | DROP	Delete the link between details stored in library and the library.
CHANGE	Change the current workspace.
TRANSFER | ABSOLUTE or RELATIVE	Transfer elements from one workspace to another.

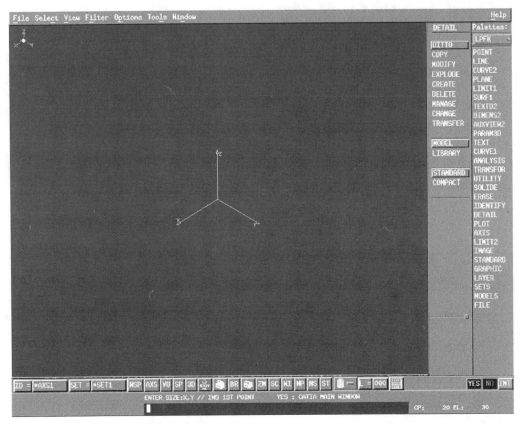

DETAIL function menu in SPACE mode.

MERGE in SPACE Mode

The MERGE function in SPACE mode is used to merge part or all elements from the "sending" model to the "receiving" model. The receiving model has been temporarily saved using the Save current model in tmp file button on the fixed menu. Options are described below.

MERGE in SPACE Mode

Option	Description
SELECT | ELEMENT	Select an element to be copied.
SELECT | VIEW	Select a view to be copied.
SELECT | SET | GEOM or SPECIFIC	Select a set to be copied.
SELECT | DETAIL or SYMBOL	Select a detail or symbol to be copied.
MERGE	Merge the selected entities into another model.

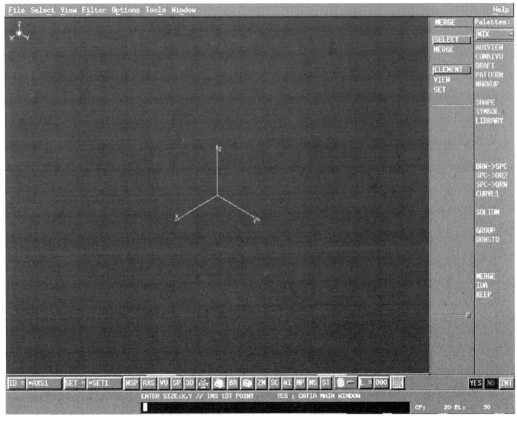

MERGE function menu in SPACE mode.

Exercise 1: Duplicating and Positioning the Side Plates

To begin, read the *BEAM TROLLEY SOLID SIDE PLATE* model saved in Chapter 11, or read it from the companion CD. The side plate solid must be correctly positioned and duplicated as presented in the following steps.

1. Create and apply a 90 degree counter-clockwise rotation about the z axis. Select TRANSFOR > CREATE | ROTATE. Key in *Z* and press <Enter>; key in *90* and press <Enter>.

   ↦ **NOTE:** *Occasionally, CATIA does not accept a key-in of the axis name (i.e., X, Y, or Z). If a message such as "Element Type Invalid" displays, select the axis instead.*

2. Select TRANSFOR > APPLY | REPLACE | FAMILY. Choose the side plate, and click on the YES button to confirm.

   ↦ **NOTE:** *By using the FAMILY option, all elements logically linked to the side plate are also transformed. Using this option is almost always recommended.*

3. Create and apply an 86mm translation along the y axis. Select TRANSFOR > CREATE | TRANSLAT. Key in *Y* and press <Enter>; key in *86* and press <Enter>.

4. Select TRANSFOR > APPLY | REPLACE | FAMILY. Select the side plate then click on the YES button to confirm.

5. Create and apply a 3.7mm (0.146") translation along the z axis. Select TRANSFOR > CREATE | TRANSLAT. Key in *3.7* and press <Enter>.

6. Select TRANSFOR > APPLY | REPLACE | FAMILY. Select the side plate and then click on the YES button to confirm.

7. Create and apply a symmetry about the xz plane. Select TRANSFOR > CREATE | SYMMETRY. Key in *XZ* and press <Enter>.

Exercise 2: Duplicating and Positioning the Wheels 269

8. Select TRANSFOR > APPLY | DUPLICATE | FAMILY. Upon selecting the side plate, a duplicate is created.

9. Select ERASE > NOSHOW. Key in *spc-*sol and press <Enter>. This is the multi-select instruction to hide (no show) all space elements except solids.

10. Save the model as *SIDE PLATES CORRECTLY POSITIONED*.

Completed beam trolley side plates.

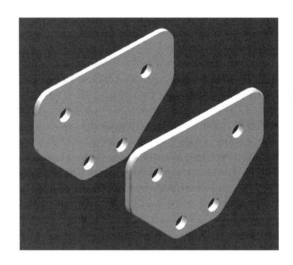

➥ **NOTE:** *The completed model of the side plates is available on the companion CD.*

Exercise 2: Duplicating and Positioning the Wheels

Read the *BEAM TROLLEY SOLID WHEEL* model saved in Chapter 11 or read it from the companion CD. Take the following steps to correctly position the wheel solid and duplicate it three times.

1. Using the same methods as in Exercise 1, create and apply the following transformations to the wheel: (a) translation along y of *-84mm* (-3.3096"); (b) translation along z of *-46.3mm* (-1.824"); and (c) translation along x of *+74mm* (+2.9156").

2. Select DETAIL > TRANSFER | ABSOLUTE. Key in *WHEEL DETAIL* and press <Enter>, and then click on the YES button to continue. Click on YES again to confirm, and then click YES again to transfer all elements into the detail workspace.

3. Select ERASE > ERASE; delete the wheel solid. Select DETAIL > DITTO | MODEL | STANDARD. Press <Enter> to select the wheel detail, and then key in ,, and press <Enter> to create the ditto. Click on the YES button to end.

4. Select DETAIL > MODIFY | TRANSLATE | DUPLICAT. Select the wheel ditto, and then the x axis. Key in *-148* and press <Enter>. A duplicate wheel ditto is created. Click on the YES button to end.

5. Select DETAIL > MODIFY | SYMMETRY | DUPLICAT. Select one of the wheel dittos. Select FLIP XZ from the menu; another duplicate will be created. Click on the YES button to end. Repeat the last step on the other wheel ditto. You should now see four wheels in respective correct positions.

6. Choose the Tools pull-down menu and then select Change Color. Click in the Color Num box and key in *95* to select a magenta color from the window. Select the four wheels in turn to change wheel color.

7. Transfer the axis to layer 10, and then save the model as *BEAM TROLLEY 4 WHEELS COMPLETE*.

Completed four wheels of beam trolley.

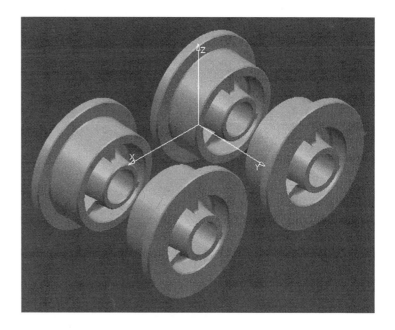

Exercise 3: Merging the Spindle Assembly Parts into the Wheel Model

For this exercise, you will use the *BEAM TROLLEY WHEEL SPINDLE ASSEMBLY* model from the companion CD. The exercise will demonstrate how elements from one model can be copied into another using the MERGE function.

1. Read the *BEAM TROLLEY 4 WHEELS COMPLETE* model saved in the previous exercise. Click on the Save current model in tmp file button on the fixed menu. The model is now temporarily saved.

   ◆◆ **NOTE:** *Remember, the correct button can be found by slowly moving the cursor along the fixed menu buttons. Brief descriptions of button functions display at the bottom left corner of the message area.*

2. Select FILE > FILE. Press <Enter> to view the list of available files and select the one into which you saved the *BEAM TROLLEY SPINDLE ASSEMBLY* model from the CD. The current option in the menu will change to READ. Read *the BEAM TROLLEY SPINDLE ASSEMBLY* model.

3. Change to the MIX palette and select MERGE > SELECT | SET. Select any element; all elements will be highlighted. Select MERGE > MERGE, and then click on the YES button to execute the merge.

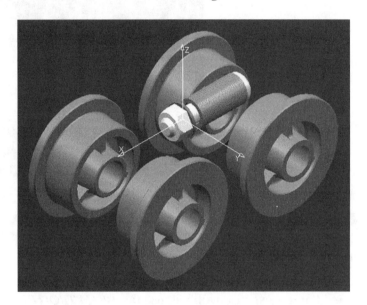

Four wheels with the spindle assembly model after the merge.

The spindle assembly solids have been merged into the four wheels model. To move the solids into the correct position, take the following steps.

1. First, a secondary axis system must be created. Select POINT > COORD | SINGLE | RECTANG. Key in *74,96,-46.3* and press <Enter>. A point will be created at the specified coordinates.

2. Select AXIS > CREATE | Z-AXIS. Select the newly created point and a highlighted 3D axis will appear. Click on the YES button to create the axis.

Exercise 3: Merging the Spindle Assembly Parts into the Wheel Model

3. To correctly orient the axis, select AXIS > SWAP, and select the x axis of the new axis system. Next, select the y axis of the same axis system and the axes will swap. Select AXIS > INVERT, and select the y axis of the new axis system. The new system will be inverted.

4. To reinstate the original axis system as the current one, click on the Change current axis system button on the fixed menu and then select the original axis. Click on the YES button.

5. Select TRANSFOR > CREATE | MOVE. Select the original axis system, and then select the new axis system that you created. The message MOVE DEFINED displays.

6. Select TRANSFOR > APPLY | REPLACE | SET. Choose any element of the spindle assembly, and then click on the YES button to confirm the move. The spindle assembly solids will move to the correct position in one of the wheels.

7. To insert the spindle assembly into the other wheels, select DETAIL > TRANSFER | ABSOLUTE. Select the original wheel ditto. Note that all switch to dimmed mode. The message FROM: *MASTER TO: WHEEL displays.

8. Key in *SET* (multi-select for set), and then select any of the spindle assembly solids. Click on the YES button to confirm the transfer, and then click on the YES button again to end. Note that each wheel now has a spindle assembly inserted in it. The spindle assembly solids were transferred into the detail workspace of the wheel, and there are four dittos of the wheel in the master workspace.

9. Select ERASE > ERASE, and key in *sol* and press <Enter> to erase solids so that only dittos remain.

10. Save the model as *BEAM TROLLEY 4 WHEELS + SPINDLES COMPLETE*.

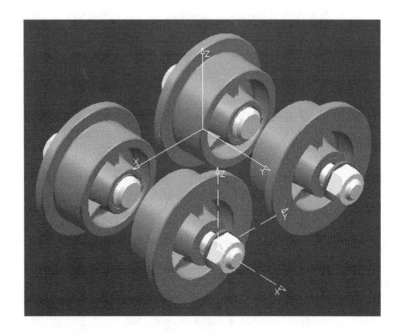

Four wheels complete with spindle assemblies.

Exercise 4: Using IMAGE to Create Windows and Screens

The objective in this exercise is to create a screen consisting of four windows showing the views on xy, yz, xz, and xyz. Take the following steps.

1. Select IMAGE > SCREEN | DEFINE. Select the bottom right STANDARD SCREEN option (four windows as described above).

Exercise 4: Using IMAGE to Create Windows and Screens 275

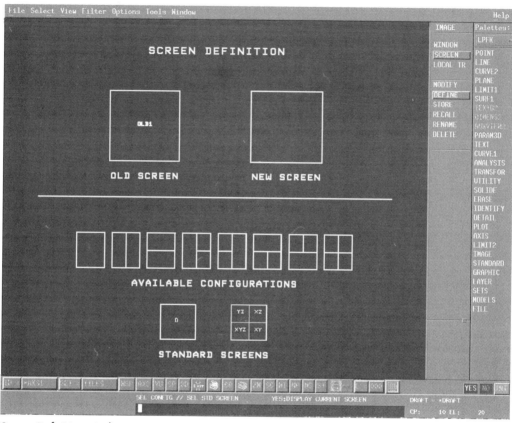

Screen Definition window.

2. Select IMAGE > LOCAL TR. Choose the axis icon in the corner of each window in turn to make it active. Use the dials (if you have access to same), arrow keys, and mouse to manipulate each window until the respective image is fully inside the boundary of the window as shown in the next illustration.

3. The screen can be stored by selecting IMAGE > SCREEN | STORE. Key in *SC4* and press <Enter>. The four-window screen will be saved as *SC4*.

Chapter 13: Managing Solids

Four-window screen created using IMAGE.

This chapter has reintroduced selected management tools available in CATIA. Using the models you have saved in this chapter, practice with other options from the function menus such as GROUP, LAYER, SETS, and LIBRARY.

Summary

Topics covered in this chapter include defining and applying transformations on solids, using details to produce duplicates, merging elements from one model to another, and creating and storing windows and screens.

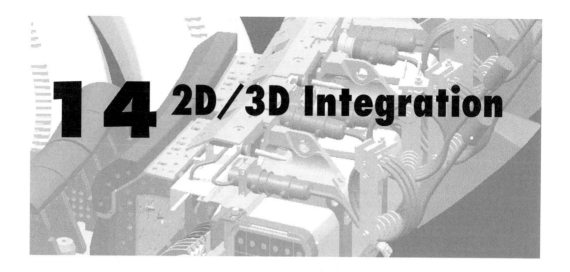

14 2D/3D Integration

In previous chapters, you worked in DRAW mode or SPACE mode. In contrast, this chapter is focused on transferring elements between DRAW and SPACE. This method of working is known as 2D/3D integration and is extremely useful when you wish to use geometry created in DRAW while working in SPACE mode to directly use as geometry or to construct solids. 2D/3D integration can also be used when a solid model assembly has been completed and you wish to produce a DRAW manufacturing assembly drawing.

Using DRW→SPC, SPC→DRW, and SPC→DR2

Functions covered in this chapter are accessed from the palette menus. The functions you use depend on what you wish to transfer and in which direction (DRAW to SPACE or SPACE to DRAW). Functions and options are described below.

Function	Description	Option	Description
DRW→SPC	SPACE mode function used in transferring elements from DRAW to SPACE.	CREATE	Create geometry in SPACE mode using elements from DRAW mode.

Base and cone.

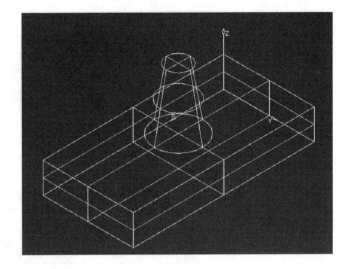

Cylinder

1. Select SOLIDE > CREATE | CANONIC | CYLINDER. For the anchor point of the solid, key in *75,37.5,75* and press <Enter>.

2. Select the Z axis for the direction. Enter the following values in the CYLINDER panel: R1 = *10* (outer radius); H1 = *50* (height); H2 = *0*; and R2 = *0*.

3. Click on the YES button to create the cylinder.

Base, cone, and cylinder.

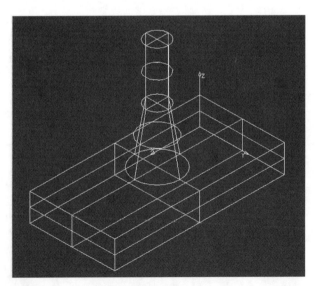

Chapter 14: 2D/3D Integration

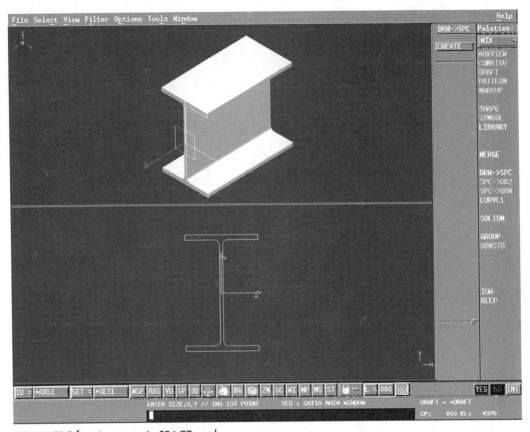

DRW→SPC function menu in SPACE mode.

Function	Description	Option	Description
SPC→DRW	DRAW mode function used in transferring elements from SPACE to DRAW.	CUT	Create geometry in DRAW mode by cutting SPACE mode wireframe and surface elements.
		PROJECT	Create geometry in DRAW mode by projecting SPACE mode wireframe and surface elements.

Using DRW→SPC, SPC→DRW, and SPC→DR2

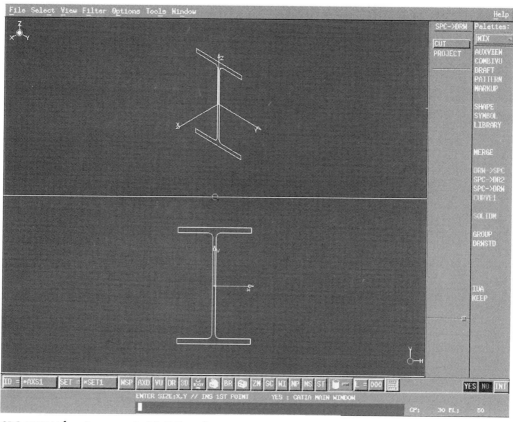

SPC→DRW function menu in DRAW mode.

Function	Description	Option	Description
SPC→DR2	DRAW mode function used in creating planar representations of solids in DRAW.	CUT	Create geometry in DRAW mode by cutting SPACE mode solids using a plane.
		PROJECT	Create geometry in DRAW mode by projecting SPACE mode solids, volumes, or set of faces onto a plane.

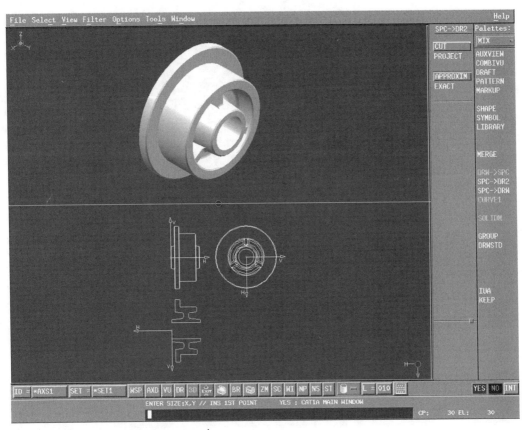

SPC→DR2 function menu in DRAW mode.

Exercises in this chapter are based on models completed in previous chapters. If you have not completed the models employed in the exercises, access the required model files from the companion CD.

Exercise 1: Transferring Elements from DRAW to SPACE

The objective of this exercise is to use geometry created in DRAW mode for the I beam to fashion a solid of the I beam. Geometry from the dimensioned drawing will not be recreated in SPACE mode.

Exercise 1: Transferring Elements from DRAW to SPACE

Read the I beam model saved in Chapter 3, Exercise 6. The screen display is in DRAW mode. A split screen display is necessary for this exercise, that is, SPACE mode in the top half and DRAW mode in the bottom half. In order to create the split screen take the following steps.

1. Select IMAGE > SCREEN | DEFINE and select the screen option that shows the screen split horizontally.

2. Define the window you wish to see in the top half of the screen. Select the window that shows XYZ. For the bottom half, select the window that shows D. Click on the YES button to confirm creation of the split screen.

3. Save the screen by selecting IMAGE > SCREEN | STORE. Key in *SC2* and press <Enter>.

The new geometry will be created in the SPACE work area. Check the SP/DR button. If the button reads DR, click on it to change it to SP.

> **NOTE:** *If you cannot see all the geometry in the DRAW screen, click on the red circle at the bottom right of the screen. You will then be able to zoom in and out using the arrow keys or the dials.*

To transfer the geometry from the DRAW work area to the SPACE work area, take the next steps.

1. To define the plane on which geometry will be placed, select PLANE > THROUGH and choose any element in the DRAW view. A plane created on the YZ axis is highlighted. Click on the YES button to confirm proper placement of the plane.

2. Select DRW → SPC > CREATE. Upon selecting the plane in SPACE, the plane on which the geometry is to be placed is defined.

3. When selecting the elements to be transferred, you have two options. First, you can individually select each of the elements in the DRAW view. Note that when an element is selected it is copied in SPACE and becomes temporarily nonselectable in the DRAW view. Second, you can multi-select all DRAW view elements by keying in *DRW* and pressing <Enter>. Once again, when the elements are selected they are copied in SPACE and become temporarily nonselectable in the DRAW view. After transferring the geometry, click on the YES button to end the selection.

The I beam geometry has been created on the SPACE plane without the necessity of using the dimensions. At this juncture, you can create a solid from the geometry by taking the following steps.

1. Select SOLIDE > CREATE | CANONIC | PRISM. To define the contour of the shape to be created, select an element from the SPACE geometry, and (to complete the contour) either click on the AUTO SEARCH button or select another element from the contour that is not adjacent to the first selected element. When you have selected the contour, click on the YES button to end the contour selection process.

2. To define the limits of the solid prism, key in *,-250* to the input information area and press <Enter>, or key in the limits to the prism limit information window. The I beam solid is created. Upon changing the display mode to shaded, the completed model should resemble the following illustration.

3. Save the model. Use a new name to avoid overwriting the original model.

Exercise 2: Creating Solids from DRAW Geometry

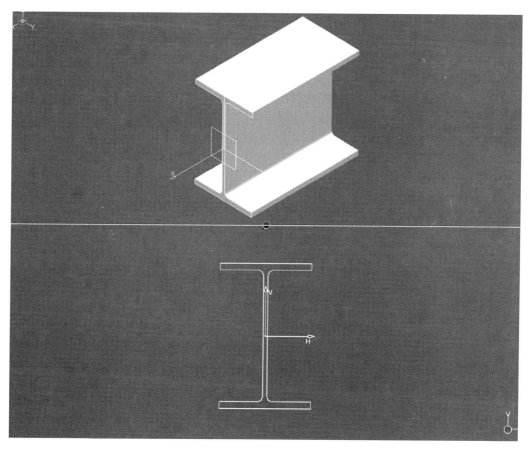

I beam solid created using DRAW geometry.

Exercise 2: Creating Solids from DRAW Geometry

The objective of this exercise is to use the geometry created in DRAW mode for the side plate to create a solid of the side plate, but without first recreating the geometry in SPACE mode from the dimensioned drawing or keying in dimensions for plate thickness.

Read the side plate model saved in Chapter 7, Exercise 1. When you read this model the screen display is a DRAW mode screen. In order to complete this exercise, a split screen is required with SPACE mode in the top half and DRAW mode in the bottom half. In order to create the split screen follow the steps listed in the previous exercise and again store the screen as *SC2*.

Verify that you are working in SPACE mode. Check the SP/DR button; if the button reads DR click on it to change it to SP.

To transfer the geometry from the DRAW work area to the SPACE work area, take the following steps.

1. To define the plane on which the geometry will be placed, select PLANE > THROUGH and select any element in the front DRAW view. A plane is created on the YZ axis; the plane is highlighted. Click on the YES button to confirm proper placement of the plane.

2. Select DRW→ SPC > CREATE and then select the plane in SPACE. The plane on which the geometry is to be placed is defined.

3. When selecting the elements to be transferred, you have two options. First, you can individually select each of the elements in the front DRAW view. Note that when an element is selected it is copied in SPACE and becomes temporarily nonselectable in the DRAW view. Second, you can multi-select all the front DRAW view elements by keying in *DRW* and pressing <Enter>. Again, when the elements are selected they are copied in SPACE and become temporarily nonselectable in the DRAW view. When all geometry is transferred, click on the YES button to end the selection.

The side plate geometry has been created on the SPACE plane without the necessity of using the dimensions. At this juncture, you can create a solid from the geometry by taking the following steps.

1. Select SOLIDE > CREATE | CANONIC | PRISM. To define the contour of the shape to be created, select an element from the SPACE geometry and either click on the AUTO SEARCH button to complete the contour. Select each of the circles in turn to create an inner profile. After selecting the outer and inner profiles, click on the YES button to end the contour selection process.

2. To define the limits of the solid prism, you need not key in the same as in the previous exercise. The information you require is available in DRAW because plate thickness is shown in the side view of the side plate. Select

Exercise 3: Creating DRAW Geometry by Transferring

in turn the two lines that define plate thickness (the lines that are 10mm apart in the side view). Click on the YES button; the side plate solid is created. Upon changing the display mode to shaded, the completed model should resemble the next illustration.

3. Save the model. Use a new name to avoid overwriting the original model.

Side plate solid created using DRAW geometry.

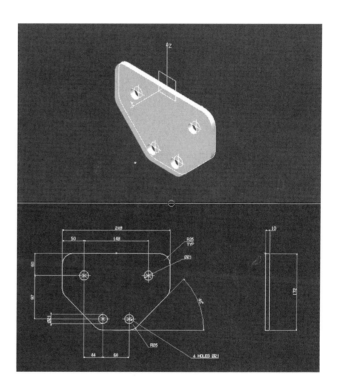

Exercise 3: Creating DRAW Geometry by Transferring from SPACE to DRAW

In Exercises 1 and 2, SPACE geometry was created by transferring from DRAW to SPACE. In this exercise, DRAW geometry is created by transferring from SPACE to DRAW.

To avoid the necessity of creating new SPACE geometry before beginning the exercise, you can use the SPACE geometry created in Exercise 1. First, however, you need to delete the DRAW geometry and the solid as demonstrated in the following steps.

1. Read the model saved in Exercise 1. To erase all the DRAW elements, select ERASE > ERASE, and multi-select the elements by using *DRW.

2. To erase the solid and the plane from SPACE, select ERASE > ERASE, and select the solid and then the plane. Alternatively, key in *SOL+*PLN.

For this exercise you will work in DRAW mode. Check the SP/DR button; if the button reads SP, change it to DR.

To transfer the geometry from the SPACE work area to the DRAW work area, take the following steps.

1. To define the view to which the geometry will be projected, select SPC→DRW > PROJECT | STANDARD or SAME, and then select the view axis. Click on the YES button to end the view selection. (If you are working with more than one view, you can project into all views selected.)

2. To select the elements to be transferred, you have two options. First, individually select each of the elements in SPACE. Note that when an element is selected it is projected into DRAW and it becomes temporarily nonselectable in SPACE. Second, you can multi-select all the SPACE elements by keying in *SPC and pressing <Enter>. Again, note that when the elements are selected they are projected into DRAW and become temporarily nonselectable in SPACE. The SPACE elements have been projected into DRAW.

3. Save the model. Use a new name to avoid overwriting the original model.

Exercise 4: Producing DRAW Views from Solids 287

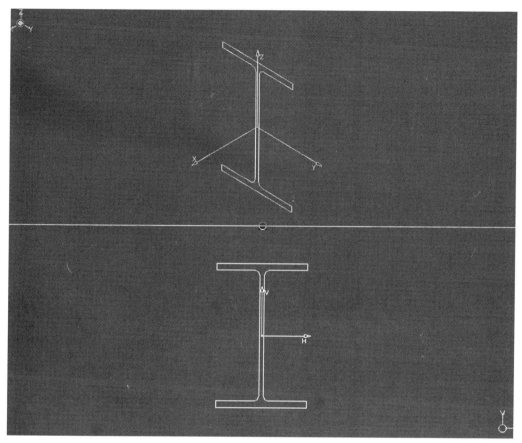

DRAW geometry created from SPACE geometry.

Exercise 4: Producing DRAW Views from Solids

The objective of this exercise is to produce DRAW views from the solid model of the wheel created in Chapter 11. Read the model of the solid wheel saved in Chapter 11, Exercise 4. The screen display is in SPACE mode. A split screen is necessary for this exercise, that is, SPACE mode in the top and DRAW mode in the bottom half. The required split screen should already be stored. Access the screen by clicking on the SC button and keying in *SC2* and pressing <Enter>. An alternative method for accesing the split screen is to select IMAGE > SCREEN | RECALL, key in *SC2*, and press <Enter>.

If you need to create the split screen, use IMAGE > SCREEN | DEFINE. (Refer to Exercise 1 as necessary.)

The wheel model should also have been saved with an applied filter. Consequently, you will not see the DRAW axis. Because you will need the axis for this exercise, select LAYER > FILTER | APPLY | DIRECT | GENERAL, and select the filter ALL from the filter window. You will now be able to view the DRAW axis.

To create the DRAW views from the solid, take the following steps.

1. Select SPC➔DR2 > PROJECT | EXACT and select the DRAW view axis. Click on the YES button to end the view selection. (If you have more than one view, you can project into all views selected.)

2. The HIDDEN LINES MODE window allows you to define how the DRAW view will appear. Set the Hidden lines option to Choose, and the Intensity option to Remove. Select the wheel solid from the SPACE window. In the DRAW view, DRAW elements have been created to represent the wheel.

> **NOTE:** *After you have completed this exercise, repeat it and change the settings in the HIDDEN LINES MODE window to view the effect on created views.*

3. Having created the first view, the plan and front views will be created in steps 3 and 4. (Refer to Chapter 6 for information on creating plan and front views.) Select AUXVIEW > CREATE | NEW BGD, and select the H (horizontal) axis. Click on the YES button to ensure that the same origin is used for the new view. Indicate a position for the new view below the existing view. Key in the following name for the new view: *PLAN*.

4. To create the front view, repeat the process in step 3. Instead of selecting H, select the V (vertical) axis of the original view. Indicate a position to the right of the original view, and name the view *FRONT*.

5. To create the side view, select SPC➔DR2 > PROJECT | EXACT. Select the front DRAW view axis. Click on the YES button to end the view selection.

6. Select the wheel solid. In the DRAW view DRAW elements have been created to represent the wheel.

At this juncture, a section through the wheel will be created. Follow the steps below.

Exercise 4: Producing DRAW Views from Solids

1. Select SPC➔DR2 > CUT | EXACT. As before, select the view in which you require the new elements. In this instance, the view is *PLAN*.

2. When asked to define the cutting plane for the section, you can click on the YES button to accept the default setting. This view was created on the XY plane.

   ➥ **NOTE:** *If you required the section to be other than on the XY plane you would have to key in a dimension from the XY plane to the position at which you required the section.*

3. Having defined the position of the section you can now select the wheel solid. Note that DRAW elements representing the section through the wheel have been created. To finish off the section you could cross hatch the section using PATTERN.

4. Save the model. Remember to use a new name to avoid overwriting the original model.

DRAW views created from SPACE solids.

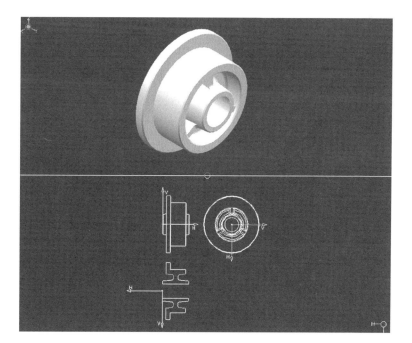

Summary

Topics covered in this chapter include transferring elements from DRAW to SPACE, transferring elements from SPACE to DRAW, and creating planar representations of solids in DRAW.

15 Working in a Multi-Model Environment

The purpose of this chapter is to demonstrate how multiple models can be superimposed on one another or overlaid, how multiple model groups can be stored as sessions, and how these models and sessions can be managed. The only new function introduced in this chapter is MODELS. The File pull-down menu will be used to read and save models and sessions.

The MODELS function is used to manage, modify, and copy elements contained in overlaid models. Options available in the MODELS function in SPACE mode are described below.

MODELS Function in SPACE Mode

Option	Description
MANAGE | ANALYZE	Analyze the status of active and passive models.
MANAGE | UPDATE	Update models contained in a session to reflect changes.
MODIFY | REPLACE or DUPLICATE | TRANSLATE or ROTATE or SYMMETRY or MOVE	Perform transformations on overlaid models.
COPY | ELEMENT or FAMILY or SET	Copy elements from a passive model to the active model.
BREAKOUT	Differentiate between the active model and duplicate occurrences.
WORKAREA | APPLY or CREATE or DELETE or MODIFY or LIST	To create and manage a subset of overlaid models.

Chapter 15: Working in a Multi-Model Environment

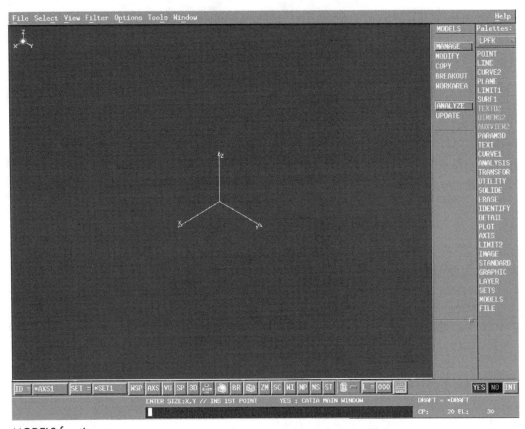

MODELS function menu.

File pull-down menu.

Exercise 1: Creating a Multi-model Environment

Before proceeding to the creation of a multi-model environment, a little setup is required. Setup steps appear below.

1. This exercise will be carried out in an empty model. Select FILE > CREATE | YZ. Verify that the SP/DR button is set to SP, and then click on the YES button. The empty model has been created.

2. Change the current layer to *010* using the Set Current Layer button on the fixed menu. Next, select LAYER > FILTER | APPLY | DIRECT | GENERAL. Select LAYCUR (layer current) from the FILTERS window so that only layer *010* is visible. This step was necessary because the beam trolley solids were saved with the solids on layer *010* and the geometry on layer *000*. When the models of the beam trolley components are overlaid, the geometry should not be visible unless you specify such visibility.

3. Verify that the display mode is set to Shaded by selecting Set current display mode and choosing the bottom (shaded) option.

Setup steps are now complete. To build the multi-model assembly of the beam trolley, follow the procedures in the rest of this section.

1. From the File pull-down menu select Open or press <Ctrl>+O. In the ensuing window, verify that the following selections are made before proceeding—Access: CATIA Declaration Files; File Format: MODEL; Open Mode: Add Passive; and Sort: Name.

2. Click on the Options button, and then click on the Keep Current Screen Layout button so that a tick is showing. Ensure that Absolute Axis is selected for both Passive and Active Models, and then click on the OK button.

3. In the list of files in the Directories box, double-click on the file in which you have the beam trolley models stored. A list of models will be displayed in the Files box at the right. Select the *BEAM TROLLEY WHEEL SPINDLE ASSEMBLY* model. The model name will appear in the selection area; click on the OK button. The spindle assembly model will be overlaid as a passive model on the empty start model.

4. Select ERASE > W.SPACE, and then select an element of the spindle assembly. The ELEMENT NOT IN THE CURRENT MODEL message is displayed. This step demonstrates that the spindle assembly model is passive.

5. From the File pull-down menu, select Open once again to return to the Open window. Note that the right angle bracket (>) symbol is displayed near the *BEAM TROLLEY WHEEL SPINDLE ASSEMBLY* model, which means that it has already been overlaid. Now select the *BEAM TROLLEY 4 WHEELS COMPLETE* model by double-clicking on it. (The alternative to double-clicking is to click on the model once and then click on OK.) The four wheels model is overlaid.

To move the spindle assembly to the correct position, take the following steps.

1. Create a point by selecting POINT > COORD | SINGLE | RECTANG. Key in *74,96,-46.3* and press <Enter>. A new point is created. To create a point on the origin, key in ,, and press <Enter>.

2. Select MODELS > MODEL | MODIFY | REPLACE | TRANSLATE, and then select an element of the spindle assembly. The whole spindle assembly model is highlighted.

To define the translation, follow the steps below.

1. Select the point on the origin, and then select the newly created point. A vector showing the defined translation and the TO INVERT: SEL ARROW // YES: CONFIRM prompt will be displayed.

2. Verify that the arrows are pointing in the correct direction, and then click on the YES button to confirm the translation.

Next, to rotate the spindle assembly, continue with the steps below.

1. Select MODELS > MODIFY | REPLACE | ROTATE, and then choose any element of the spindle assembly model.

2. Select the Z axis of the highlighted local axis of the spindle assembly for the rotation axis. Key in *90* and press <Enter>, and then click on the YES button to complete the rotation. The spindle assembly is now in the correct position in relation to the first wheel.

Exercise 1: Creating a Multi-model Environment

To duplicate the spindle assembly into the next wheel, take the next steps.

1. Select MODELS > MODIFY | TRANSLATE | DUPLICATE, and then select any part of the spindle assembly.

2. Select the Y axis of the highlighted axis system. Key in *148* and press <Enter>.

3. Click on the YES button to confirm the translation. A duplicate spindle assembly is created in the next wheel.

To duplicate the spindle assemblies into the other two wheels, take the following steps.

1. Select MODELS > MODIFY | DUPLICATE | SYMMETRY, and then select MULTI-SEL from the MODEL SELECTION window.

2. Select any element in each of the two spindle assemblies. Click on the YES button to end the selection.

3. Key in *XZ* for the plane of symmetry, and then click on the YES button to duplicate the spindle assembly into the other two wheels. The four wheels are now complete with respective spindle assemblies.

Four wheels complete with spindle assemblies.

In the following steps, the side plates are added to the assembly.

1. From the File pull-down menu, select Open.
2. Select the *BEAM TROLLEY SIDE PLATES COMPLETE* model, and click on the OK button. The side plates model is overlaid.

To add the stud assembly to the overall assembly, continue with the steps below.

1. From the File pull-down menu select Open.
2. Select the *BEAM TROLLEY ASSEMBLY* model from the companion CD, and click on the OK button. The first stud assembly is overlaid.

To duplicate the stud assembly in the other position, follow the next steps.

1. Select MODELS > MODIFY | DUPLICATE | TRANSLAT.
2. Choose an element of the stud assembly. Select the X axis of the main axis system, key in *60*, and press <Enter>.
3. Verify that the arrow is pointing in the correct direction. If it is not, select the arrow and it will invert. Click on the YES button to confirm. The stud assembly is duplicated into the other position.

The last part of the beam trolley assembly is the I beam.

1. From the File pull-down menu select Open.
2. Select the *BEAM TROLLEY SOLID I BEAM* model. Click on the OK button, and the I beam will be overlaid.

To ensure that only the solids are visible, follow the next steps.

1. Select ERASE > NO SHOW | W.SPACE.
2. From the Select pull-down menu, choose Space Axes. The space axes will be transferred to No Show.
3. From the Select pull-down menu, select Basics to hide the points. Only the solids are visible.

To save the overlaid models as a session, proceed as described below.

1. From the File pull-down menu, select Save Session As.

Exercise 1: Creating a Multi-model Environment

2. From the Directories box, double-click on a session file to which you have access. (Check with your system administrator about file access.)

3. Click in the Selection box and key in *BEAM TROLLEY FINAL ASSEMBLY*.

4. In the Save Sessions Options window, click on the Save References Only button, and then click on the OK button. The session will be saved with each of the models as it currently exists in the file (i.e., Save references only).

5. To view the status of each model in the session, select MODELS > MANAGE | ANALYZE. An ANALYSIS window displays as shown in the next illustration.

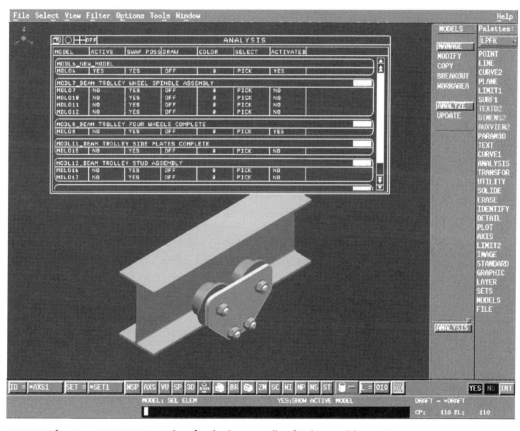

MODELS function ANALYSIS window for the beam trolley final assembly session.

Exercise 2: Copying Elements Between Models

The objective of this exercise is to copy one of the solid wheels of the beam trolley into a new model. Setup steps appear below.

1. Select FILE > CREATE | YZ. Verify that the SP/DR button on the fixed menu reads SP. Click on the YES button to create a new empty model.

2. To overlay the solid wheels model, select Open from the File pull-down menu. Change File Format to MODEL. Verify that Open Mode is set to Add Passive. Double-click on the file in the Directories box which contains the *BEAM TROLLEY 4 WHEELS COMPLETE* model, and then double-click on that model in the Files box. The wheels will be overlaid on the empty model.

To copy a wheel from the passive model to the active model, proceed with the following steps.

1. Select MODELS > COPY | ELEMENT, and then select one of the wheels. Click on the YES button to end, and then click on the YES button again to confirm the copy. The selected wheel has been copied to the active model.

2. To close the passive model, select Close from the File pull-down menu. Select the *BEAM TROLLEY 4 WHEELS COMPLETE* model from the Close window and click on the OK button. The passive model will be closed leaving the copied wheel.

➢ **NOTE:** *This is an extremely useful method for bringing elements previously created into a new model, particularly because they are copied in the exact position in which they occur in the passive model.*

Exercise 3: Using BREAKOUT

Assume that you require a second I beam which must resemble the original, but with additional features or modifications. Take the following steps after reading the *BEAM TROLLEY SOLID I BEAM* model.

Exercise 3: Using BREAKOUT

1. Select MODELS > MODIFY | DUPLICAT | TRANSLAT and choose the I beam solid.

2. Key in *X* and press <Enter>, and then key in *650* and press <Enter>. Click on the YES button to confirm. A duplicate occurrence of the I beam model is created.

3. Select MODELS > BREAKOUT. Key in *second I beam* and press <Enter>. The BREAKOUT DONE message is displayed.

4. Select MODELS > MANAGE | ANALYZE. Note that two models are displayed.

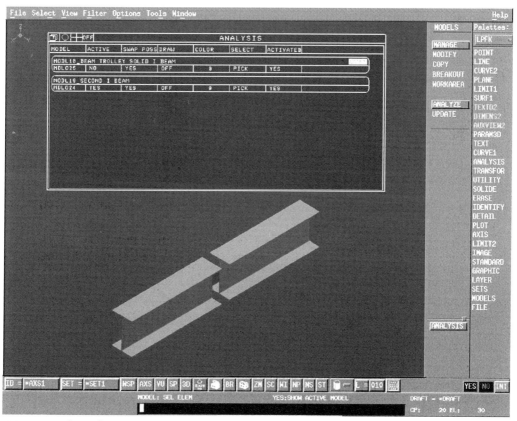

MODELS analysis after BREAKOUT.

With the use of BREAKOUT, the new model has become the active model. Its links with the original are broken and modifications can be carried out.

Exercise 4: Creating Work Areas

The objective of this exercise is to create separate work areas within the multi-model environment in which different models can be active. After reading the *BEAM TROLLEY FINAL ASSEMBLY* session, take the following steps.

1. Select MODELS > WORKAREA | CREATE | EMPTY. Key in *I Beam* and press <Enter>. Select the I beam, and then click on the YES button to end.

2. Key in *Side Plate* and press <Enter>, and then select a side plate. Click on the YES button to end.

3. Repeat the above procedure to create a work area for the wheels.

4. To view the results of step 1, select MODELS > WORKAREA | APPLY. Press <Enter> to view the list of work areas, and select *I_Beam* from the list. You are now in the *I_Beam* work area where *BEAM TROLLEY SOLID I BEAM* is the active model.

5. Using the described method, apply one of the other work areas. To determine whether the model is active, simply select FILE > WRITE. The name of the active model will appear in the information entry area.

When using work areas, it is possible to perform modifications on individual models in discrete areas and then save a session with a different name to record the changes. To demonstrate this procedure, take the following steps.

1. Apply the *I_Beam* work area. Change the color of the solid to green using the GRAPHIC function or the Change Color option from the Tools pull-down menu. Apply the *Side_Plates* work area and change color to red. Apply the *Wheels* work area and change color to blue.

2. Reapply the Default Work area to view the changes you have made. From the File pull-down menu, select Save Session As. Double-click on the session file in the Directories box. In the Selection area, key in *Colored Beam Trolley* and press <Enter>.

3. In the Save Session Options window, click on the Save copies of all data button, and then click on OK. The new session is saved.

Exercise 4: Creating Work Areas 301

Save Session Options window.

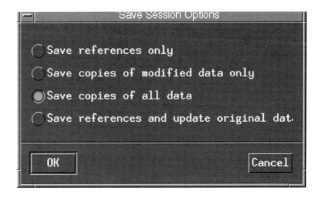

4. To clear the screen, select FILE > CREATE | YZ, and then click on the YES button.

5. From the File pull-down menu, select Open. In the Open window, verify that File Format is set to SESSION and double-click on the *COLORED BEAM TROLLEY* session which will be read. Note that the new colors are now in force.

6. Using the method described above, read the *BEAM TROLLEY FINAL ASSEMBLY* session. Click on the OK button in the Warning window that appears. The session is read showing the original colors.

In the original session, the models were saved using the Save References Only option, which simply stores pointers to the models as they exist in the file. Every time the session is read, the models are read from file. In the colored session, the models were saved using the Save Copies of All Data option so that a copy of each model reflecting changes (such as the changes in color) is stored.

You are encouraged to practice building sessions, copying and dropping models, creating work areas, and so forth with any of the models saved on the companion CD. Experiment with different ways of using the techniques explored in this chapter. All models required to create the *BEAM TROLLEY FINAL ASSEMBLY* can be read from the companion CD.

Summary

This chapter focused on creating a multi-model environment, manipulating and managing models within a multi-model environment, and storing multi-model environments as sessions.

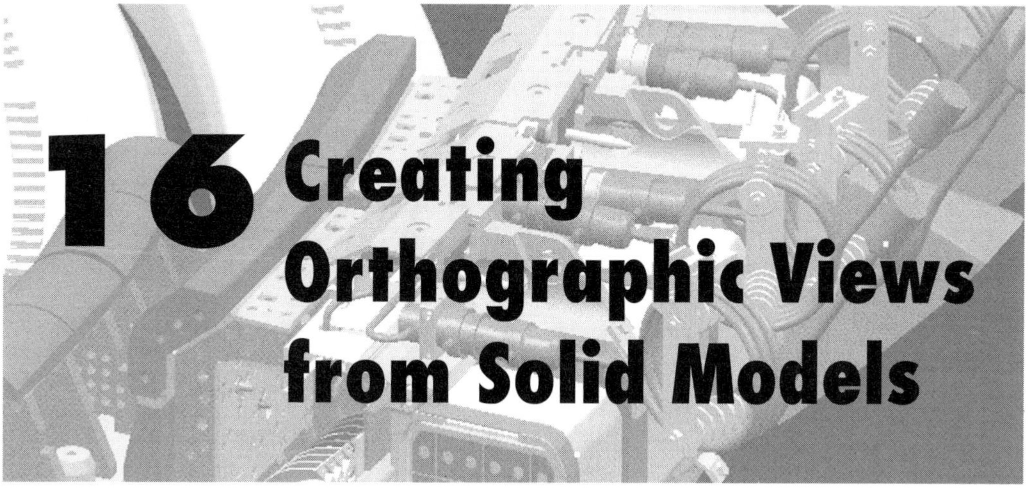

16 Creating Orthographic Views from Solid Models

Chapter 14 covered the use of 2D/3D integration to create elements in DRAW views from existing SPACE elements and SPACE elements from existing DRAW elements (i.e., static one-time projection). In contrast, this chapter is focused on creating orthographic views and sections from SPACE solids. Upon using AUXVIEW2, when changes are made to solids in a model the views previously created from the solids, as well as dimensions, can be updated (i.e., dynamic projection). The following functions will be either introduced or developed further in the current chapter.

- AUXVIEW2
- TEXT
- AUXVIEW

AUXVIEW2 options are numerous. The discussion below covers types of views that can be created and the most commonly used AUXVIEW2 options.

AUXVIEW2

The AUXVIEW2 function is used to create and modify views. As seen in the next illustration, the AUXVIEW2 menu structure differs from most other functions in

CATIA in that the options are not selected from the side menu but from the Objects and Action screen windows.

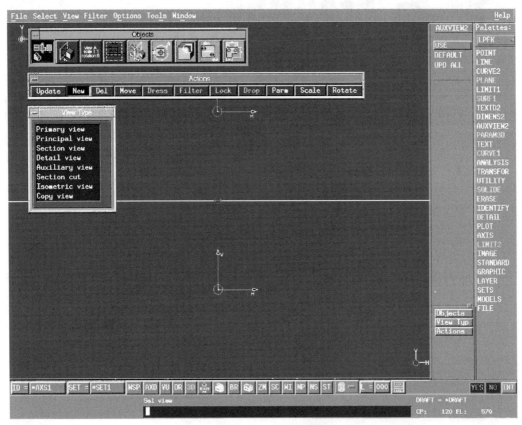

AUXVIEW2 function.

Objects window (Advanced mode).

Action windfow.

While the side menu and Action window options are single words, the Objects window options are displayed as icons. Appearing below is an explanation of the icons and a single word definition that will be used in subsequent lists for selectable options. The captions reproduce text in the message area that appears when passing the cursor over the respective icons.

View. Manage the view entity.

Plane. Define a new current view projection.

Text. Manage the text associated with a view.

Frame. Manage a view frame.

Back_pln. Perform a back clipping.

Clip. Manage a view clipping.

Breakout. Manage a simple or (unspec) view.

Section. Manage a view graphical and mathematical.

Dimens. Manage dimensions.

In the following tables, the first option appearing in the left column is selected from the side menu; the second from the Objects window, and third and subsequent options from the Action window.

Options	Description
View options created with AUXVIEW2	
USE	VIEW	NEW	PRIMARY VIEW	Primary views from which other orthographic views and sections can be created.
USE	VIEW	NEW	PRINCIPAL VIEW	Principal views or views defined by orthographic projection.

Options	Description
USE I VIEW I NEW I SECTION VIEW	Sectional views showing geometry beyond the cutting plane.
USE I VIEW I NEW I DETAIL VIEW	Detail view (enlarged view of small area, such as O ring grooves).
USE I VIEW I NEW I AUXILIARY	Auxiliary view (defined by lines not normal or parallel to the view axis).
USE I VIEW I NEW I SECTION CUT	Section cut views not showing geometry beyond the cutting plane.
USE I VIEW I NEW I ISOMETRIC	Isometric views.
USE I VIEW I NEW I COPY VIEW	Copy view (duplicate existing view).

Modifying views

USE I VIEW I UPDATE	Update a single view (when a solid in the model is modified).
UPD ALL	Update all views.
USE I VIEW I DEL	Delete view.
USE I VIEW I MOVE	Move view.
USE I VIEW I DRESS	Modify graphical attributes of DRAW elements in a view.
USE I VIEW I FILTER	Filter out solids in a view.
USE I VIEW I FILTER	Uncut solids in a section view.
USE I VIEW I LOCK	Lock view; modification is possible only through AUXVIEW2 options.
USE I VIEW I DROP	Isolate view to prevent update when solids change.
USE I VIEW I PARM	Change general parameters of view.
USE I VIEW I SCALE	Modify view scale.
USE I VIEW I ROTATE	Rotate view.
USE I TEXT	Modify text associated with a view.
USE I BACK_PLN	Modify depth of viewing beyond cutting plane with section views and cuts.
USE I CLIP	Clip existing views to remove unwanted elements.
USE I BREAKOUT	Create part sections in existing views.
USE I SECTION	Modify cutting planes in existing sections.
USE I DIMENS	Move dimensions created automatically from parameterized solids.

Parameters for creation of views are defined with the use of the AUXVIEW2 function's DEFAULT option.

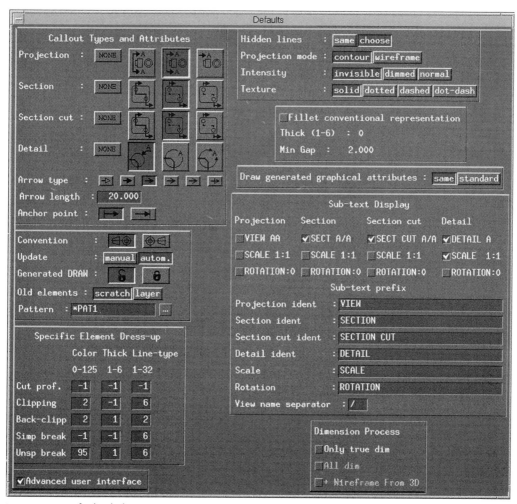

AUXVIEW2 Defaults dialog.

The Defaults panel is organized into groups of options as described below.

- **Callout types and attributes**. Define how section lines, projection lines, and arrows are displayed.

- **Hidden lines, projection mode, intensity and texture**. Define how created elements appear in the view. For instance, intensity and texture allow the user to show hidden lines in various line types.

- **Fillet conventional representation**. Modifies the display for fillets created in SOLIDE > OPERATE | FILLET.

- **Draw generated graphical attributes**. Allows attributes from the SPACE elements or the DRAW standard to be used for the creation of new views.

- **Sub-text display**. Defines the text displayed and the wording used for views and sections.

- **Dimensional process**. Controls automatic dimensioning from parameterized solid models.

- **Advanced user interface**. Controls options available in the Objects window.

- **Specific element dress-up**. Defines the graphical representation of section and clipping lines.

The last options panel contains the selections described below.

- **Convention**. Defines first or third angle projection.

- **Update**. Create a view showing only SPACE elements (Manual) or DRAW elements (Autom).

- **Generated DRAW**. Lock the view on creation. A locked view can be modified only by using AUXVIEW2 options.

- **Old elements**. Automatically erase old elements after a view update (scratch) or place old elements on a chosen layer (layer).

- **Pattern**. Defines the pattern (hatching) applied to sections.

➙ **NOTE:** *Prior to performing the exercises in this chapter, the Default panel should be set as shown in the previous illustration. Upon completion of the exercises, we recommend that you modify the default settings in order to fully understand their use.*

TEXT

TEXT function options in SPACE mode are summarized in the following table.

Option	Description
CREATE	Create text associated with SPACE elements.
MODIFY	Modify the attributes of SPACE text.
ERASE	Erase SPACE text.
MOVE	Modify the anchor point of SPACE text.
EDIT	Modify SPACE text.
SHOW	Modify the visibility of SPACE text.

Unlike DRAW text, SPACE text is attached to an existing element. The text may contain the elements identifier or keyed information.

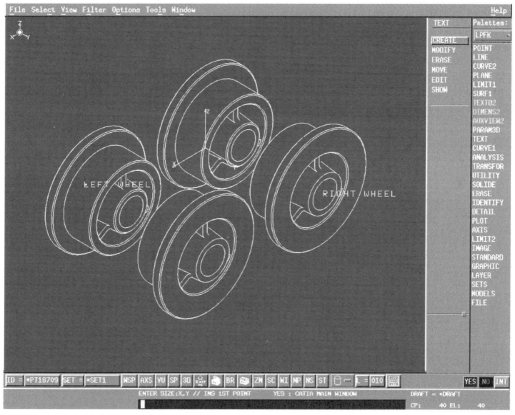

TEXT function in SPACE mode.

AUXVIEW

With the AUXVIEW function, you can create views without DRAW elements (DRAW views of the SPACE model). This feature can be useful if a simple picture of a solid assembly drawing is required for ballooning purposes.

The following exercises employ models created in previous chapters. If you have not completed the required exercise from a previous chapter, you can read the model from the companion CD.

Exercise 1: Using AUXVIEW

The objective of this exercise is to create DRAW views containing no DRAW elements. First, read the following model saved in Chapter 13, Exercise 2: *BEAM TROLLEY 4 WHEELS COMPLETE*.

The model should be viewed with a split screen in order to see both the SPACE and DRAW workspaces. If a split screen has already been created in the model, you can change the screen display in either of the following ways: (1) click on SC button, key in *SC2*, and press <Enter>, or (2) select IMAGE > SCREEN | RECALL, key in *SC2*, and press <Enter>. If a split screen is not available, take the following steps. (Refer to Chapters 1 and 15 for further information.)

1. Select IMAGE > SCREEN | STORE and save the SPACE screen as *SC3*.

2. Define a split screen using IMAGE > SCREEN | DEFINE as shown in the next illustration. The upper half of the screen is the SPACE work area and the lower half, the DRAW work area. The only element in the DRAW screen is an axis. Save the split screen as *SC2* using IMAGE > SCREEN | STORE.

3. This model may also have been filed with a filter applied. A filter is not necessary in this exercise. Select LAYER > FILTER | APPLY | DIRECT | GENERAL; from the filter window select filter ALL.

In the DRAW work area part of the screen you will see only an axis system. Take the following steps to create the end view.

1. Click on the SP button in the fixed menu area to change it to DR.

Exercise 1: Using AUXVIEW

2. Select AUXVIEW > MODIFY | SPACE and select the view, or click on the YES button because the view you wish to modify is the current view. The SPACE elements now appear in DRAW view because all DRAW views created are planar views of the SPACE model. In this instance, the default view was created on the YZ plane.

> *NOTE: The current view contains only SPACE elements. Upon using the ERASE function and accessing an element in DRAW or SPACE modes, you will see that the element disappears from DRAW and SPACE. Do not forget to click on NO during the last operation; otherwise, you will have to start the exercise from scratch.*

If you now wish to create another orthographic view of the SPACE model that contains only SPACE elements, use the following steps or refer to Chapter 6.

1. Select AUXVIEW > CREATE | NEW BGD, and then select the V axis. Click on the YES button to accept the same origin.

2. Indicate the position for the new view in the DRAW window. Key in a name for the new view. Note that you have created a second view with only the SPACE elements showing. If you do not require the SPACE elements to be visible in a particular view, you would use AUXVIEW > MODIFY | SPACE and select that view.

Chapter 16: Creating Orthographic Views

Orthographic and isometric views of the beam trolley four wheels assembly.

The final segment of this exercise is to create an isometric view of the solid assembly. In order to create this view use the following steps.

1. Select AUXVIEW > CREATE | NEW BGD. Indicate in the SPACE window, or key in *XYZ*.

2. Indicate the position for the new view in the DRAW window. Key in a name for the new view.

You have created three views in the DRAW work area without having created any DRAW elements. This maneuver would be beneficial for an assembly drawing where only the annotation need be added in DRAW.

Exercise 1: Using AUXVIEW 313

Instead of creating text in DRAW mode you could now add some text to the wheel solids, and the text would also appear in the DRAW views. To create the SPACE text as shown in the next illustration, take the steps listed below. First, toggle the SP/DR button to SP.

SPACE text labeled solids.

1. Text in SPACE mode must be attached to an existing SPACE element. Consequently, create two points by selecting POINT > COORD | SINGLE | RECTANG.

2. For the first point, key in *74,-84,-46.3* and press <Enter>. This point will be created on the center line of the left wheel. For the second point, key in *-74,84,-46.3*. These points have not been created in the detail workspace because SPACE text cannot be attached to elements within a detail.

3. Select TEXT > CREATE and then select the first point. The text created will be attached to this point. In the TEXT DEFINITION window, choose the MANUAL TEXT option which allows you to key in the required text. If the ELEMENT ID option is active, the text to be created will be the element identifier. Set text size to NORMAL. In the TEXT panel, key in *LEFT WHEEL*. (The maximum number of characters for the TEXT panel is 70.)

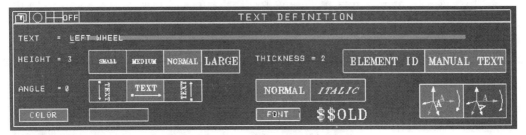

TEXT DEFINITION window.

4. Select the point again to position the created text. SPACE text has now been linked to the point on the left wheel. The text on the right wheel can now be created by repeating the previous steps; select the point on the right wheel and key in *RIGHT WHEEL* instead.

5. Save the model. Change the model name to avoid overwriting the original.

Note that the SPACE text appears in SPACE as well as in the three DRAW views. Text in SPACE always appears parallel to the screen. To verify, rotate the SPACE model using the 3D Transformation options in the Display and Manipulation tools accessed by clicking button 3 of a three-button mouse or button 4 of a four-button mouse.

SPACE text can be used to label SPACE elements. For instance, assume that you have created solids on different layers. You could identify the solids according to layer by giving each solid a piece of text containing the layer number. When working in a multi-model environment, you could label solids with respective model names in order to identify solids according to model.

Exercise 2: Using AUXVIEW2

In the previous exercise, views were created with only the SPACE elements visible. In the current exercise, views from the solid assembly will be created with DRAW elements. After creating orthographic views, the exercise continues with creating sections and detail views. Setup steps appear below.

1. Read the *BEAM TROLLEY FINAL ASSEMBLY* session saved in Chapter 15.

Exercise 2: Using AUXVIEW2 315

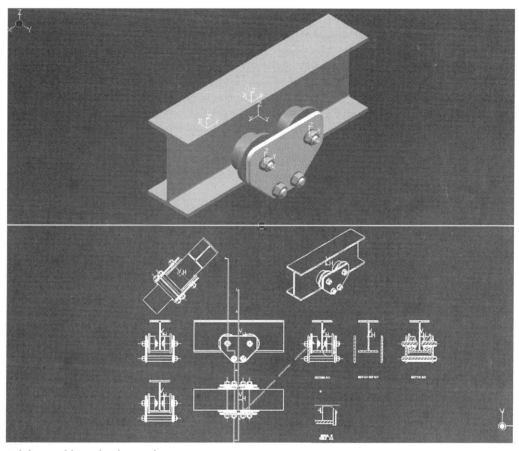

Solid assembly and orthographic views.

   2. Create a split screen in order to view both the SPACE and DRAW workspaces. (Refer to Chapters 1 and 15 for further information on creating a split screen.) Before creating the split screen, save the SPACE screen as *SC3* using IMAGE > SCREEN | STORE.

   3. Define a split screen as shown in the next illustration using IMAGE > SCREEN | DEFINE. The upper half of the screen is the SPACE work area, and the lower half, the DRAW work area. The DRAW area contains only an axis.

   4. Save the split screen as *SC2* using IMAGE > SCREEN | STORE.

316 Chapter 16: Creating Orthographic Views

➡ **NOTE:** *Before proceeding with this exercise, verify that the SP/DR button reads DR and that the AUXVIEW2 default panels are set as described at the beginning of this chapter.*

To create views from the solid assembly with DRAW elements, follow the steps below.

1. The first view to be created is the primary. Select AUXVIEW2 > USE | VIEW | NEW | PRIMARY VIEW.

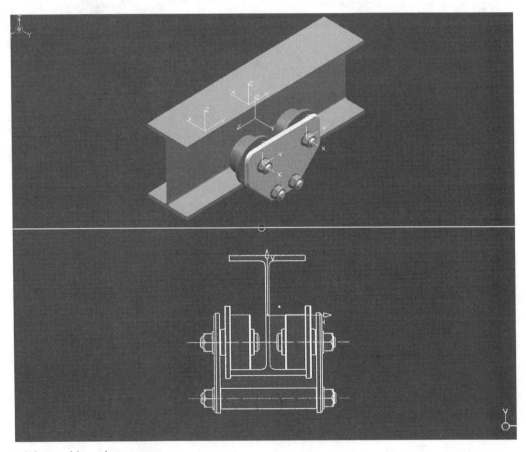

Solid assembly and primary view.

2. When asked to define the origin and viewing plane, select the Y axis and then the Z axis from the central current axis system in the SPACE work area.

Exercise 2: Using AUXVIEW2

3. Next, indicate a position in the DRAW work area when asked to define the position of the view. An end elevation, the first view created using AUXVIEW2, is complete. AUXVIEW2 names the view automatically, although you can rename it if you desire. Note that the origin of the new view occurs at the intersection of the two selected lines from the SPACE work area (the Y and Z axes).

Once the view has been created, the Action window option changes to MOVE. This option allows you to move the position of the view. If you indicate a new position, the center of the view will move to that point.

The next views to be created are principals, or side elevation and plan. These views are defined by orthographic projection. The principal views created in the next series of steps are shown in the following illustration.

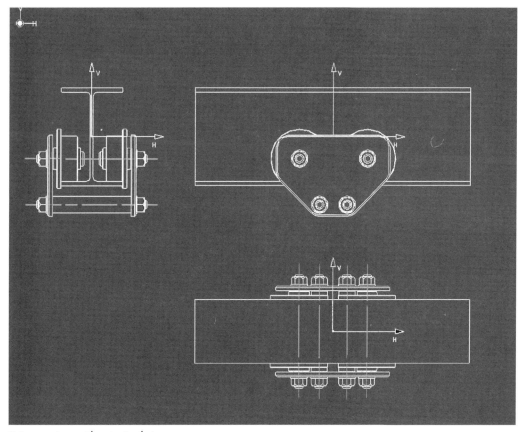

Primary view with principal views.

Before proceeding, change the screen display to the DRAW work area only. Define the DRAW screen by using the IMAGE function, or click on the WI button in the fixed menu area and key in *D*.

1. To create the side elevation view, select AUXVIEW2 > USE | VIEW | NEW | PRINCIPAL VIEW. At this point, the screen is divided by two diagonal highlighted lines. Indicate in the right quadrant; the side elevation is created.

2. The plan view is created by using the side elevation view rather than the end elevation as the defining view. Select AUXVIEW2 > USE | VIEW | NEW | PRINCIPAL VIEW. Choose the side elevation view to change the defining view. The intersection point for the diagonal lines will be the axis of the side elevation view. Indicate in the bottom quadrant; the plan view is created.

The next in line to create is a section view.

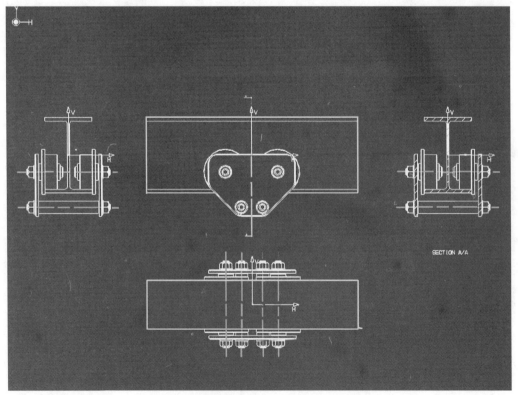

Section view added.

Exercise 2: Using AUXVIEW2 319

1. Select AUXVIEW2 > USE | VIEW | NEW | SECTION VIEW. The side view should be the current view. In AUXVIEW2, the current view is indicated by a highlighted circle on the axis. If the side view is not the current view, select it.

2. Select the V axis in the side view. A section line with two vector arrows normal to it appears. These vectors define the viewing direction for the section. If they are pointing the wrong way, select one of the vectors to reverse it. Click on the YES button to accept the vector direction shown.

3. Indicate a position for the view; the section is created.

   NOTE: *A section view displays all elements that lie beyond the cutting plane as well as on the plane.*

The next view to create is a detail view.

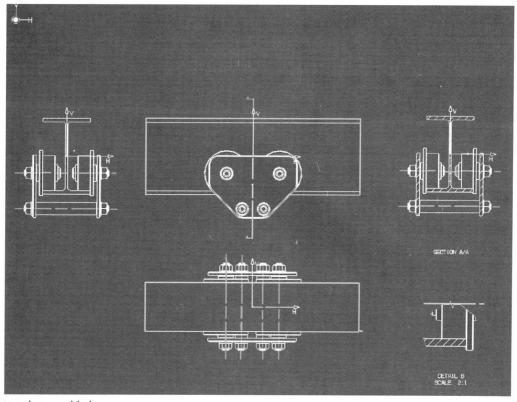

Detail view added.

320 Chapter 16: Creating Orthographic Views

1. Select AUXVIEW2 > USE | VIEW | NEW | DETAIL VIEW. Use the section view, which is the current view, as the defining view.

2. Indicate two points in the area where the wheel touches the I beam. These points should lie on the circumference of the defining circle and in opposition to one another. Click on the YES button to accept the profile of the detail view.

3. When asked to enter a value for the view scale, key in *2* for a scale of 2:1. (If no value is entered, the default value will be used.)

4. Indicate a position for the view; the detail view is created.

The next view to create is an auxiliary view.

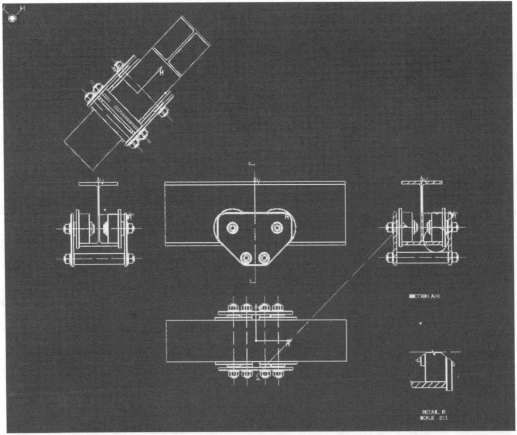

Auxiliary view added.

Exercise 2: Using AUXVIEW2

1. Select AUXVIEW2 > USE | VIEW | NEW | AUXILIARY VIEW. Select the side view as the defining view. Choose the angled line at the left to define the viewing line. The viewing line will be normal to the selected line.

2. Indicate a position for the view to define the side from which the view is taken. The auxiliary view is created.

The next view to create is a section cut.

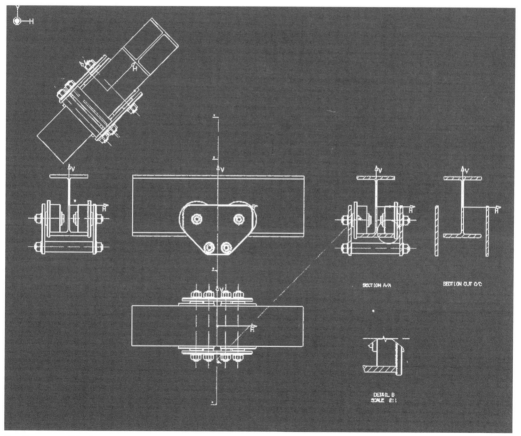

Section cut added.

1. Select AUXVIEW2 > USE | VIEW | NEW | SECTION CUT. The side view is the current view. In AUXVIEW2 the current view is indicated by a highlighted circle on the axis. If the side view is not the current view, select it.

2. Select the V axis in the side view. A section line with two vector arrows normal to it appears. These vectors define the viewing direction for the section. If they are pointing the wrong way, select one of the vectors to reverse it. Click on the YES button to accept the vector direction shown.

3. Indicate a position for the view; the section is created. With a section cut you will only see elements that lie on the cutting plane, not beyond it.

The next view to create is an isometric view.

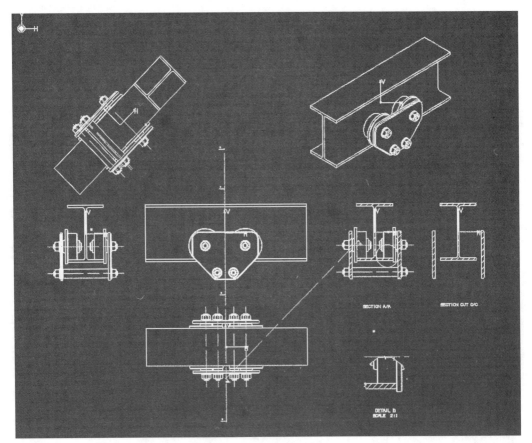

Isometric view added.

1. Select AUXVIEW2 > USE | VIEW | NEW | ISOMETRIC VIEW. Switch the screen display back to the split screen: click on the SC button on the fixed menu and key in *SC2*.

Exercise 2: Using AUXVIEW2

2. Change the display mode by clicking on the Set current display mode button and selecting the hidden line removal mode. Select any element in the SPACE window to define a parallel window plane.

3. Indicate a position for the view in the DRAW window; the isometric view is created.

The last new view to create is a copy of an existing view.

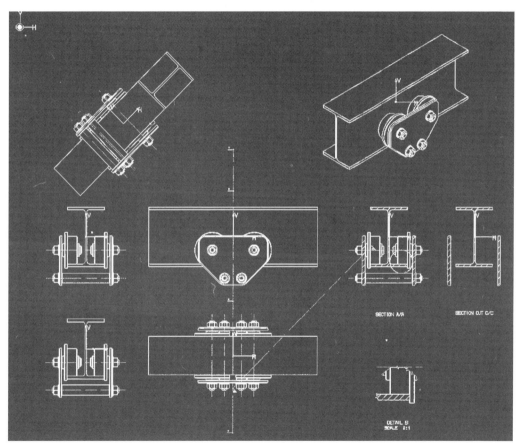

Copy of end view added.

Select AUXVIEW2 > USE | VIEW | NEW | COPY VIEW. To copy the current view, indicate a position for the new view and the view will be copied. To copy a view other than the current view, select the view to be copied and indicate a position for the new view.

You have now created all available view types. However, before moving on to modifying any of the created views, a staggered section will be created through the center line of the wheel and the center line of the retaining studs.

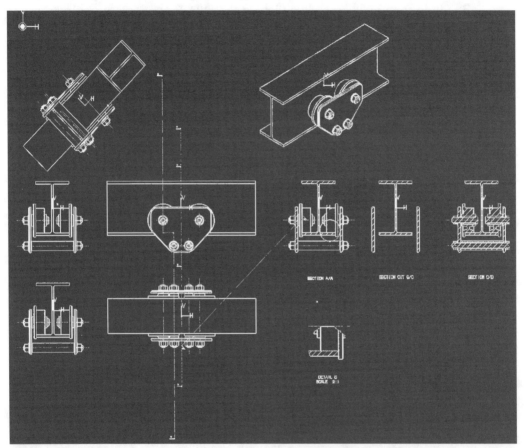

Staggered section added.

1. Create center lines on the washers in the side elevation view, the second view that was created. Select MARK UP > AXIS | LINEAR. Select the diameter relating to the washers in four positions to create the center lines.

2. Select AUXVIEW2 > USE | VIEW | NEW | SECTION VIEW. Because the section will be created from the side elevation view, select the vertical center line on the upper left washer of the wheel retaining nut, and then select the vertical center line on the lower left washer of the retaining studs.

3. Verify that arrows are pointing in the right direction. When arrow direction is correct, click on the YES button to accept the section line.

4. Indicate a position for the view; the staggered section view is created.

NOTE: *The above steps are identical when working with a section cut view.*

Note that all the hatching is at the same angle. Obviously, the hatching on adjacent components should be at different angles. You could delete all the hatching and create more by using the PATTERN function. The difficulty with this strategy is that if you update the view the hatching will revert back to its original state. A better solution is to modify the hatching using PATTERN > VISUALTN | MODAL | GRAPHISM | VIEW.

Select the hatching you wish to change. In the ensuing window, change the angle option by keying in *90* and pressing <Enter>. The angle of the selected hatching changes. You can repeat this operation by selecting additional hatching and changing its angle. When hatching is revised in this way, the new angle information is retained when a view is updated.

Now that all views have been created, save the active model with a new name, such as *TROLLEY ASSEMBLY VIEWS*. After saving the active model, save the session with a new name (*TROLLEY ASSEMBLY WITH AUXVIEWS*). When saving the session, use the option model references only.

Exercise 3: Updating Views

The UPDATE option allows you to update views when modifications are made to the solid models. When updating you have two options: (a) updating all views in the model or (b) updating individual views. Using the session saved from Exercise 2, follow the steps below to modify the solid model assembly and then update the views.

1. Select ERASE > NO SHOW, and then select the beam solid.

2. Select AUXVIEW2 > USE | VIEW | UPDATE. With this update option, you can proceed one view at a time. Select the side view and click on the YES button to accept the update. The view is updated and the beam will no longer be visible in the side view, but will continue visible in all other views.

3. To update all views, select AUXVIEW2 > UPD ALL. Click on the YES button. At this point, all views will be updated and the beam will not be visible in any of them.

4. To return the beam to all views, select ERASE > NO SHOW, swap into the NO SHOW area, and select the beam. Swap back to the SHOW area. Select AUXVIEW2 > UPD ALL and click on the YES button to update all views. The beam is once again visible in all views.

Updating works in the same way after a solid has been erased, new solids have been created, models have been added to the session, or when solids are modified in any way. Therefore, views can very simply be kept up to date whenever changes take place in the model. Next, AUXVIEW2 works equally well on a session, an assembly created in a single model, and a single solid in a model.

Exercise 4: Modifying Views

Filtering Solids in a View

The FILTER option in AUXVIEW2 allows you to filter out solids from particular views or to uncut solids that have been sectioned in a section view or cut. The first FILTER option to explore is Un_cut, used for removing sectioning from items such as bolts, nuts, shafts, and washers, and so forth.

To experiment with the Un_cut option, make the staggered section view the current view. If necessary, click on the VU button and select the staggered section view to make it current.

Exercise 4: Modifying Views 327

Staggered section view before using the Un_cut FILTER option.

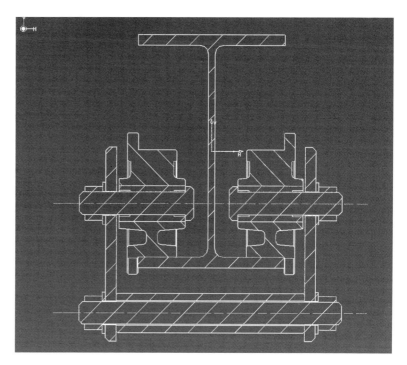

1. Select AUXVIEW2 > USE | VIEW | FILTER | UN_CUT.

2. Select the solids that you wish to uncut. One method for accomplishing this is to select an element of the solid from the section view. However, every time an element is selected the view is automatically updated.

3. Another means of selecting the required solids is to select them from the SPACE work area. Switch the screen display to a split screen by clicking on the SC button and keying in *SC2*. You could select the solids individually or globally via a multi-select option. Key in *SEL* to the input information area and press <Enter>. Select the required solids—the nuts, washers, and shafts. When you have selected all the required solids, click on the YES button to end the selection. The staggered section view should now resemble the following illustration.

Staggered section view with uncut washers, nuts, and shafts.

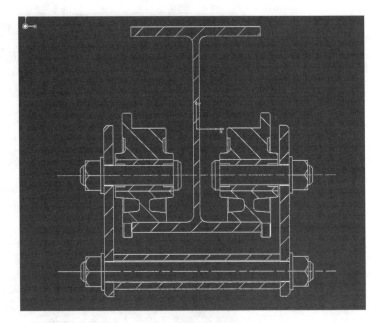

Assume that you subsequently decide to section the solids. In this instance, you would use the Re_cut option from FILTER, and follow steps 1 through 3.

The Un_use option is used for completely removing solids from views. In the steps below, use the staggered section view once more.

1. Select AUXVIEW2 > USE | VIEW | FILTER | UN_USE.

2. Select the solid or solids that you wish to unuse. This selection can be carried out in the same way as for the Un_cut option. For the moment, select an element of the I beam from the staggered view. The view automatically updates and should resemble the following illustration.

Exercise 4: Modifying Views 329

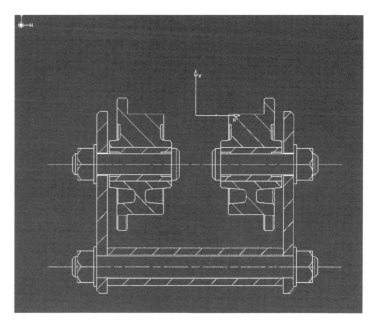

Staggered section with I beam removed.

➥ **NOTE:** *When using FILTER options on solids that are part of a detail, you will be able to select them only from the SPACE work area. In addition, the operation will be performed on all solids contained in the detail.*

Modifying the Visualization of Solids

The DRESS option is used to change the visualization of solids within a view, such as color, line thickness, line type, selectability (Pick/No Pick), and visibility (Show/No Show). You could of course use the GRAPHICS function to change any of these features, but if you do so you would have to individually select all elements associated with a particular solid. The following method is much faster.

In this exercise, the isometric view is used to modify visualization of the I beam.

1. Make the isometric view the current view. Select AUXVIEW2 > USE | VIEW | DRESS.

Chapter 16: Creating Orthographic Views

2. Select the line type option from the attribute window accessed when DRESS is selected. Change the *1* to a *2* to change the line type from a solid to dashed line. The numbers used here relate to the line type numbers used in GRAPHIC and STANDARD.

3. Select an element of the I beam from the isometric view. Every element used to show the I beam in this view changes to dashed line type as shown in the next illustration.

I beam with line type changed.

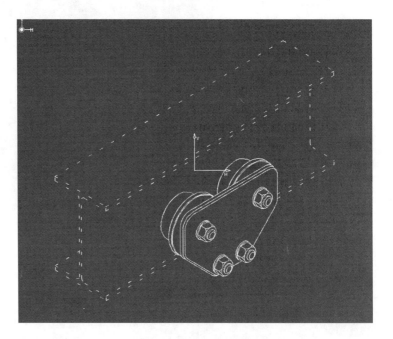

You can change other attributes of any element in the same way. Simply select the relevant option from the attribute window.

Clipping a View

The CLIPPING option is used to clip a view. Clipping is similar to creating a DETAIL view, except that the existing view is changed instead of a new view created. Use one of the end views in the model for this exercise.

1. Select AUXVIEW2 > USE | CLIPPING | ADD. Select one of the end views.

Exercise 4: Modifying Views 331

2. Indicate a series of points, such as when using a trap option, to box part of the view that you wish to retain. See the next illustration for guidance. Click on the YES button to end the definition of the clipping profile. The view is clipped and should resemble the illustration.

A clipped view.

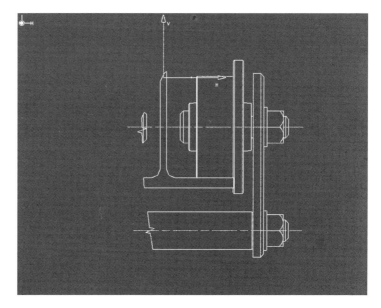

Breaking Out Solids from within a View

The next option to examine is BREAKOUT. This option is used to break away part of a solid or solids to visualize details that would otherwise be hidden. For this exercise, use the side elevation views.

1. Make the side elevation view the current view. Select AUXVIEW2 > USE | BREAKOUT | ADD.

2. Indicate a series of points, such as when using a trap option, to box part of the view that you wish to break out. Check the next illustration for guidance. Click on the YES button to end the definition of the breakout profile.

3. Note that a temporary end view has been created. From this temporary view, select the two lines that define the thickness of the side plate or indicate very close to them. The side plate will be broken away in the side elevation to reveal the wheel behind it, as shown in the following illustration.

Side view with the side plate broken away.

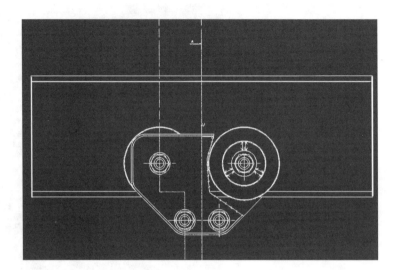

4. Save the model.

An alternative method of defining the depth of the breakout is to use planes from the SPACE model that define the thickness of the side plate. The planes could be created on each face of the relevant side plate solid by taking the following steps: if you are not already in the split screen mode, switch the display to SC2; change to SP mode; and with PLANE > THROUGH, select lines or functional surfaces.

> **NOTE**: *The side plate to be cut away is the plate on the far side of the SPACE model.*

Having created the planes you can switch back to DR mode and follow the above steps 1 and 2. The depth of the breakout could then be defined by selecting the two planes just created.

As seen in this chapter, AUXVIEW2 is a powerful function for producing many different types of DRAW views. When using modification options such as FILTER or BREAKOUT, you can use the functions together or more than once in any view. For instance, after you create a particular breakout, you may wish to create another breakout from a solid that has become visible because of the first breakout. You may also wish to use the FILTER option to create a complex break-away section.

> **NOTE:** *All of the AUXVIEW2 exercises were completed using a session and with more than one model. The same exercises could also have been completed using a single model with all solids contained in that model.*

Summary

This chapter focused on creating views that visualize the SPACE work area without creating DRAW elements; creating diverse types of views linked to the SPACE work area; and modifying views in different ways.

Before moving on to the next chapter, you are encouraged to experiment with AUXVIEW2 in the following ways: alter the AUXVIEW2 default parameters and repeat some of the exercises to observe the effect on the created views; explore modification options not covered in this chapter; and experiment with TEXT by changing the settings in the TEXT DEFINITION window.

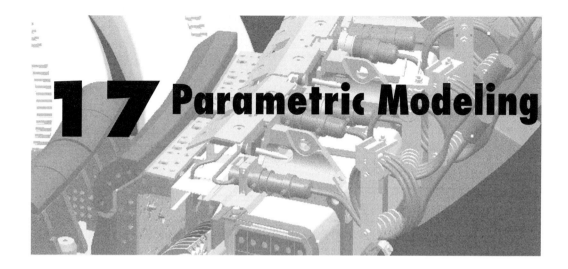

17 Parametric Modeling

The objective of this chapter is to provide an introduction to parametric modeling, where parametric dimensions are applied to 3D elements. The topics listed below are covered in the chapter.

- PARAM3D function
- Creating and managing parameterized dimensions with the SOLIDE function
- Dynamic Sketcher

The following table describes options available in the PARAM3D function in SPACE mode. The PARAM3D function in SPACE mode is used to create, apply, and manage parameterized dimensions on elements.

PARAM3D Function in SPACE Mode

Option	Description
PROFILE | CONSTRN	Define constraints on a profile.
PROFILE | CREATE	Create a profile from geometric elements.
PRIMITIV	Parameterize a primitive solid or wireframe.
LINK	Replace a reference with a parametric relationship.
UNLINK	Reverse the LINK operation.

Chapter 17: Parametric Modeling

Option	Description
MODIFY	Modify previously created parametric dimensions.
PARAMETER I ASSIGN or ISOLATE or CREATE or EDIT or DELETE	Manage parameters.
TRANSFOR I LOCAL or SET	Apply transformations to parameterized elements.
DIMENSN I SHOW or NO SHOW or RESTORE or DELETE or POSTN or VALUE	Manage parametric dimensions.
ANALYZE I RELATION or STATUS or CONSTR I PARENT or CHILDREN or DOWN STRM	Analyze parameterized geometry.
DELETE I PROFILE or PRIMITIV or RELATION or ALL	Delete some or all parameters.
ADVANCED I RELATION or COMPUTE	Define special advanced relationships.

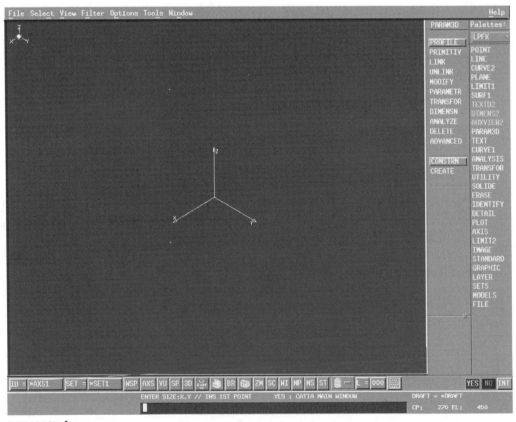

PARAM3D function menu.

Exercise 1: Creating a Parameterized Profile of Beam Trolley I Beam

To begin this exercise, read the *BEAM TROLLEY SOLID I BEAM* model saved in Chapter 11, or the file of the same name from the companion CD.

1. To ensure that the geometry is available, use the Set current layer button on the fixed menu and switch to layer *000*.

2. Select LAYER > FILTER | APPLY | DIRECT | GENERAL, and then choose the All filter from the FILTERS window. At this point, you should be able to view the solid I beam geometry with accompanying geometry and axis. The next few steps are required to parameterize the profile.

3. In this particular case, using the I beam center lines in the profile is preferable. Enter the 2D plane of the geometry by clicking on the Swap between 3D/2D button on the fixed menu. Select two elements of the I beam geometry.

4. To create the center lines, select LINE > HORIZONT | UNLIM and press <Enter> to produce an unlimited horizontal line through the origin. Use the same method to produce a vertical center line. Swap back to 3D.

5. Select PARAM3D > PROFILE | CONSTRN and then select the I beam solid. The I beam geometry will be highlighted. To include the newly created center lines in the profile, select each of them now, and then click on the YES button to end the selection.

338 *Chapter 17: Parametric Modeling*

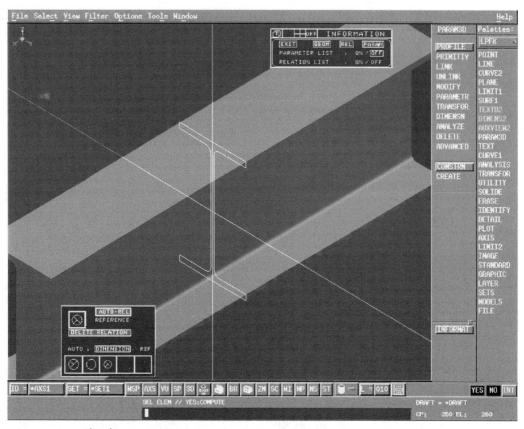

Parameterizing the I beam geometry.

6. From the window displaying the parameterization icons, select the References icon at the left of the selection box, and then successively select each of the center lines. Two reference indicators will appear on the selected lines.

7. At this point, you need to add the offset dimensions. Select the Offset icon. (When the correct icon has been selected, OFFSET will appear alongside the selected icon.)

8. Select the vertical center line, and then select one side of the web geometry. A parametric dimension with a value of 3.150 will be created.

9. To create the dimension on the other side of the center line, select the dimension you just created, and then select the vertical center line. Next, select the other side of the web geometry and indicate a position where you would like the dimension displayed.

Exercise 1: Creating a Parameterized Profile of Beam Trolley I Beam

> **NOTE:** *By selecting the dimension before selecting the geometry in this way, both the distances will be driven by one dimension.*

Using the same principle described above, create offset dimensions for the rest of the cross section geometry.

1. Select the radius icon. Upon verifying that the RADIUS option is selected, choose one of the corner radii. A 7.6 dimension appears.

2. Select the dimension and then choose the next corner. Repeat this procedure for the other two corners.

3. Click on the YES button to compute. Note that certain additional parametric dimensions will be added automatically. The parameterized profile should now resemble the next illustration.

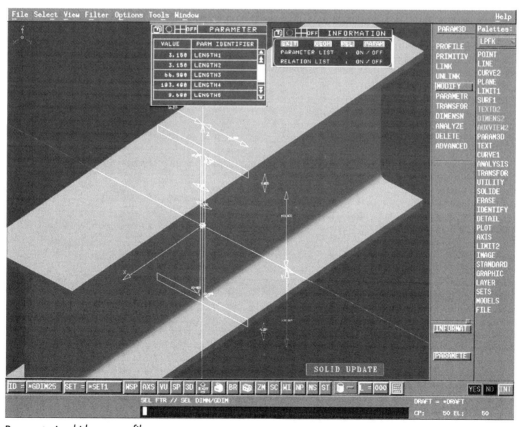

Parameterized I beam profile.

Once parameters have been created, they can be used to modify the solid. The following steps demonstrate how to modify the I beam section so that it is 250 x 150.

1. Select PARAM3D > MODIFY. From the PARAMETER window or the work area, select the *66.9* dimension (both dimensions will be highlighted).

2. Key in *150/2* (in the absence of all other digits) and press <Enter>. The dimension will change to *75*.

3. To change the height of the I beam from *206.8* to *250*, select one of the *103.4* dimensions, key in *125*, and press <Enter>.

4. Click on the YES button to accept the modifications. Next, click on the SOLID UPDATE button to update the solid. The solid will change shape according to the modified dimensions.

5. Save the model as *MODIFIED I BEAM SOLID*.

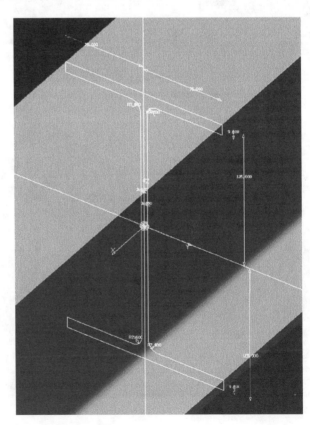

Modified I beam.

Exercise 2: Extracting DRAW Views with Autodimensioning

The objective of this exercise is to demonstrate how DRAW views can be created from the I beam solid with parameters that automatically create the dimensions. Continue using the *MODIFIED I BEAM SOLID* model for the next exercise.

1. Using IMAGE function and the method described in Chapter 13, create a screen split horizontally with two windows (a 3D XYZ window at the top and a DRAW 2D window at the bottom).

   ◆◆ **NOTE:** *If you are using Version4 R1.8, you can do this quickly by selecting Horizontal Split DR & SP from the View pull-down menu.*

2. Change to DR mode by clicking on the SPACE/DRAW toggle button on the fixed menu.

3. Select AUXIEW2 > DEFAULT. At the bottom right corner of the defaults window, verify that a tick is showing in proximity to each of the options in Dimension Process box. If ticks are not showing, select the buttons and ticks will appear. Next, at the bottom left corner verify that the Advanced User Interface is selected.

4. Select AUXVIEW > USE. Create a new primary view on the YZ plane by selecting the horizontal center line and then the vertical center line. (Use the method described in Chapter 16.) Key in a comma and press <Enter> to create the new view on top of the default view axis.

Chapter 17: Parametric Modeling

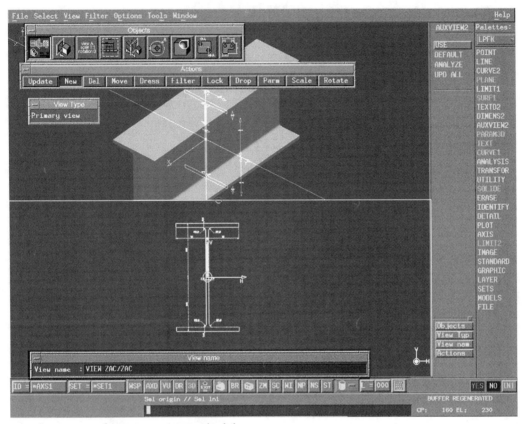

Autodimensioning from a parameterized solid.

5. A DRAW view of the I beam section is created with dimensions. If the dimensions are not ideally positioned, they can be moved using the DIMENS2 function.

The dimensions in the newly created view can now be used as parametric dimensions to modify the solid. The steps to achieve the same follow.

1. Toggle back to SP mode and select PARAM3D > MODIFY.

2. Select one of the 75 dimensions in the DRAW view, key in *66.9*, and press <Enter>. Select one of the *125* dimensions again in the DRAW view, key in *103.4*, and press <Enter>.

3. Click on the YES button to modify, and then click on SOLID UPDATE to update the solid to its original dimensions. Note that the DRAW view dimensions remain highlighted but unchanged.

4. Toggle back to DR mode and select AUXVIEW2 > UPD ALL (update all), and then click on the YES button to extract the current model. The DRAW view will be updated to match the modified solid.

5. Save the model so that you can return to it later for practice purposes.

Exercises 1 and 2 demonstrate 2D/3D integration. Drawings can be quickly produced from solid models using existing associativity between DRAW views and solids, and changes can be driven from either the solid model or the drawing.

Exercise 3: Using Parametric Dimensions to Position Features

The objective of this exercise is to demonstrate how to parameterize a solid in order to control features and profiles by parametric dimensions.

1. Read the *BEAM TROLLEY SOLID SIDE PLATE* model saved in Chapter 11, or access the file from the companion CD.

2. Change to layer *000* and apply the All filter to display all geometry.

3. Select PARAM3D > PROFILE | CONSTRN, and then select the side plate solid. Click on the YES button to end the selection. The parameter icons window appears.

4. Because the profile will not be modified, parametric dimensions for the same are not important. Therefore, you can permit CATIA to automatically generate profile dimensions by clicking on the YES button.

5. Select PARAM3D > PRIMITIV. Verify that DIMENSION is selected in the AUTO REL window, and then select the solid. A parametric dimension of the thickness of the plate is created.

Side plate after profile and primitive parameterization.

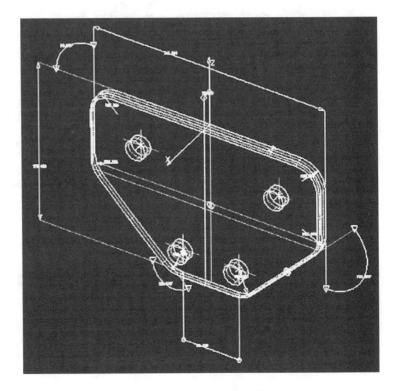

The next series of steps involves controlling the position of one of the holes by using parametric dimensions.

1. Select PARAM3D > PRIMITIV, and then select the top right hole. Parametric dimensions for the hole plus a reference indicator are created.

2. Select PARAM3D > LINK. Before proceeding, we recommend that you work in wireframe display mode in order to make the FSUR lines of the solid selectable.

3. Select the newly created reference indicator of the hole, and then choose an FSUR of an edge of the side plate from which you want to position the hole.

4. Select another FSUR on a second edge to fully define the hole position. The LINK COMPLETED: REFERENCE REPLACED message displays.

Exercise 3: Using Parametric Dimensions to Position Features 345

Parametric dimensions for hole position.

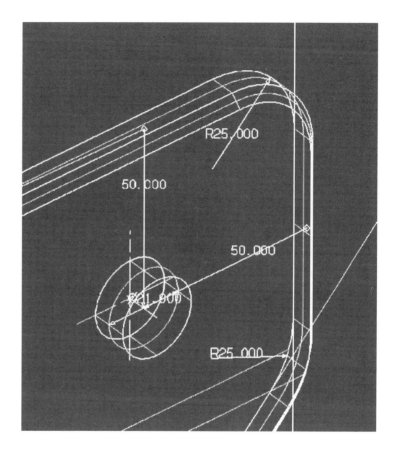

The newly created parametric dimensions of the hole can now be modified to move the hole.

1. Select PARAM3D > MODIFY, and then select each of the hole position dimensions in turn.

2. Key in *70* and press <Enter>.

3. Click on the YES button to modify, and then click on the SOLID UPDATE button to update the solid according to the modified dimensions.

Chapter 17: Parametric Modeling

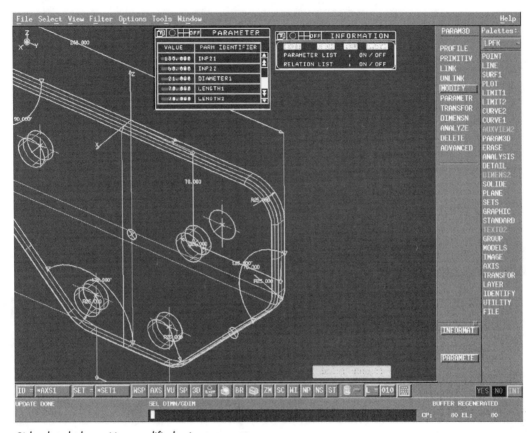

Side plate hole position modified using parameters.

Steps to perform parameterization through the SOLID functions follow.

1. Read the *BEAM TROLLEY SOLID SIDE PLATE* model and reveal the geometry as before.

2. Select SOLIDE > MODIFY | GEOMETRY | CONTOUR, and then select the solid.

3. Click on the YES button to end the selection. Note that the Parameter icon window appears; from this point on you can proceed to parameterize the contour in the same way as demonstrated above with PARAM3D. This is known as modifying parameters on the fly. As soon as you select another function, the temporary parametric dimensions will disappear and would have to be recreated for purposes of creating additional changes.

Only a few of the most commonly used facilities within the PARAM3D function have been demonstrated here. We encourage you to practice and experiment with the facilities; simply read the exercise models from the CD files. Do not be afraid to try something new because you can always start afresh.

Note on Dynamic Sketcher

For CATIA users running Version 4.1.7 or later, consider using the Dynamic Sketcher, another tool for creating and modifying geometry and solid contours. The tool is accessed by selecting Sketcher from the Tools pull-down menu. A step-by-step tutorial is available in the online CATIA documentation available in Frameviewer (accessed by double-clicking on the CatHelp icon in the CATIA_V4 window on the desktop).

In essence, the Dynamic Sketcher is a separate environment where solid contours can be modified or geometry created to bring together existing CATIA tools in a single separate area. This tool differs from other CATIA tools and functions. Experiment with it and see what you think.

> **NOTE:** *If you are about to attempt something new, and you are not sure of the result, click on the Save current model in tmp file button first. If things go wrong, you can return to the point of saving by clicking on the Restore saved model button.*

Summary

This chapter covered the following topics: parameterizing the contour of solids and primitive solids, extracting a drawing with automatically generated dimensions from a solid model, modifying a solid using parameters, and creating and modifying parameters through the SOLID functions.

18 Housekeeping Tips

Why do models require cleanup after or during creation? During model design, users typically make changes and retain the original information in case they need it later. Users may also create details that are no longer required at model completion. While all such unnecessary information may be hidden, it takes up drive space. Next, errors can occur during model creation. Errors are caused by the way in which a model is created or by software defects. Housekeeping, or the removal of unnecessary information and error resolution, is recommended both during and after model design. The functions and general command listed below are introduced in this chapter.

- IDENTIFY
- KEEP
- /CLN (CATIA general command)

In addition to the new functions and command, options to the functions listed below that were not covered in previous chapters are explored in the following exercises.

- ERASE
- DETAIL
- SYMBOL

350 Chapter 18: Housekeeping Tips

The IDENTIFY function is used to display, modify, and update element identifiers.

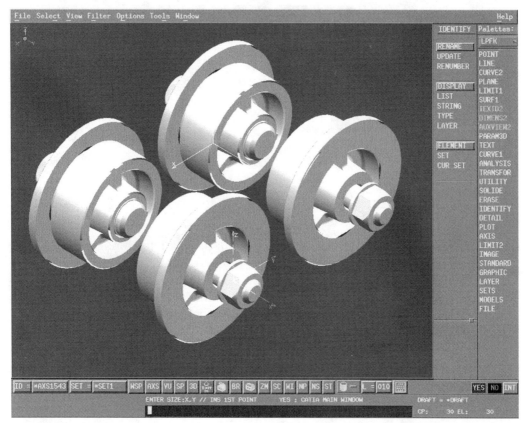

IDENTIFY function.

The KEEP function is available in DRAW and SPACE modes. Options are described below.

KEEP Function Options

Option	Description
MERGE	Merge a passive model into the active model.
SELECT	Select elements, sets, details, and applications to be retained.
KEEP	Delete all elements, sets, details, or applications not previously selected using SELECT from the model.

Function Options

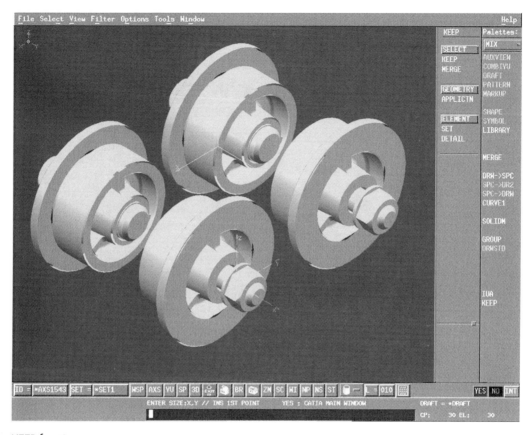

KEEP function.

/CLN is a CATIA general command which is available for use regardless of which function or mode you are working in. The /CLN command is used to (1) analyze the model for errors; (2) diagnose possible causes of errors; and (3) correct errors.

➥ **NOTE:** *Additional general commands are described in Appendix C.*

The /CLN command is not accessed via the palettes, but rather by keying /CLN to the input information area and pressing <Enter>. This command can be accessed regardless of the function being used at the time. When /CLN is entered, the panel in the next illustration appears on the screen.

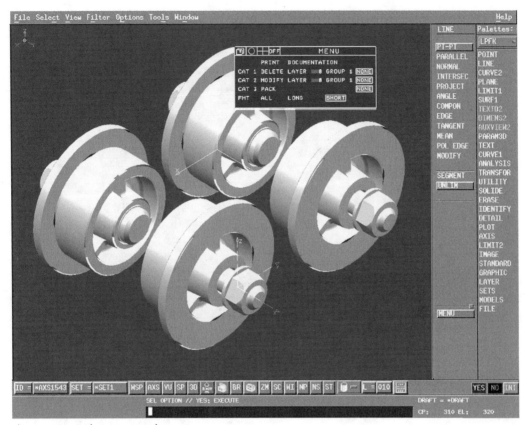

/CLN command options window.

Exercises in this chapter do not involve the creation of new models. Instead you will read some of the models created throughout the book, remove unwanted information, and check for errors in the models.

Exercise 1: Using the ERASE Function

After reading in a model, you will use the ERASE function to delete unwanted information in the NO SHOW area. Take the following steps.

1. With the FILE > READ command, read a model used in a previous chapter.

2. Select ERASE > NO SHOW. Click on the YES button to swap into the NO SHOW area.

3. If the information in the NO SHOW area is no longer required it can now be erased by selecting ERASE without swapping out of the NO SHOW area. You have the choice of individually selecting elements to erase, or you can erase them by using a multi-select option such as *DRW (to erase all DRAW elements) and *SPC (to erase all SPACE elements).

4. Having erased all the elements that you wish to erase you can swap back to the SHOW area by selecting SHOW and clicking on the YES button.

> **NOTE:** *Elements in NO SHOW linked to elements in SHOW will not be erased. For example, elements in NO SHOW used to create a solid will not be erased.*

If you have erased elements using the above method, you can check the impact on model size by selecting ERASE > PACK. The Model Status dialog appears, providing information about the size of the model. Click on the YES button to pack the model. If you have erased elements you will notice that the numbers in the PACK window have changed.

Erasing Unnecessary Details or Symbols

To eliminate superfluous details and symbols at model completion, take the following steps.

1. Select DETAIL > DELETE | UNUSED.

2. The first option is to click on the YES button to display all unused details. Having selected the detail to delete you will be asked to click on the YES button to confirm the deletion.

Using the above steps, you can delete only one unused detail at a time. If you wish to delete multiple details, you would follow the next set of steps.

1. Select DETAIL > DELETE | UNUSED | DIRECT.

2. Select details from the displayed list that you wish to delete, or choose the Select All option to delete all displayed details. Click on the YES button to confirm deletion.

The other option available to you in DETAIL > DELETE is deleting used details. This option deletes the detail and every occurrence (i.e., all dittos) of the detail in the model. To delete unused symbols, the above steps can be followed. However, you do not have to use the DIRECT | UNUSED | DIRECT option to delete symbols.

Similar to checking the impact on model size after erasing elements using NO SHOW, you can check the outcome after deleting details and/or symbols by selecting ERASE > PACK. Click on the YES button to pack the model. After erasing elements, you will notice that the numbers in the PACK window are different.

Exercise 2: Exploring the /CLN Command

During model creation, you may inadvertently generate errors. In addition, errors in the model can also be created by software defects. All such errors can be analyzed and corrected by using the general /CLN command. To analyze and correct errors, take the following steps.

1. Key in /CLN to the input information area and press <Enter>.

2. Click on the YES button, and the model will be analyzed.

3. After the model is analyzed, the results are displayed as shown in the following illustration. If errors are present, and you wish to see the full analysis, select the ALL option from the /CLN window, and then select page 2 from the Results window. At this point, you will see a description of errors, where the errors occur, and how many times they occur.

Exercise 2: Exploring the /CLN Command

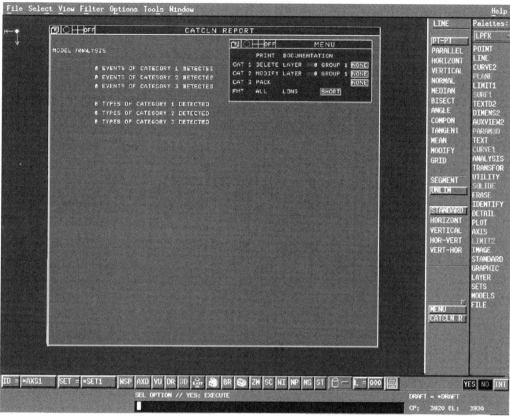

Page 1 of /CLN analysis window.

4. To remove errors from the model, select the following options from the /CLN window: DELETE (Category 1); MODIFY (Category 2); and PACK (Category 3). Click on the YES button and the error count should change to zero.

> **NOTE:** *You may need to click on the YES button more than once before the error count changes to zero.*

Similar to checking the impact on model size in the previous examples, you can check the outcome after deleting errors by selecting ERASE > PACK. Click on the YES button to pack the model. After erasing elements, you will notice that the numbers in the PACK window are different.

Exercise 3: Exploring the KEEP Function

One of the uses of the KEEP function is to delete all unwanted information for a model after first selecting the elements, sets, or details that you wish to store. To experiment with KEEP, read a model in which you have used details and take the following steps.

1. Select KEEP > SELECT | GEOMETRY | DETAIL.

2. Select a detail from the screen display; the detail will be highlighted. Choose KEEP > KEEP.

3. Click on the YES button. The only element remaining on the screen and in the model will be the detail in step 2 and element(s) logically linked to it selected in step 2. Information in NO SHOW and stored transformations, windows, and screens will also be deleted.

This example shows how to store a detail, but the same steps could be used to store any element, such as sets, views, and so forth. However, there are rules which apply to elements to be stored. These rules are summarized in the following list.

- All elements logically linked to the selected element will be stored.
- The set to which the selected element belongs will be stored.
- The view to which the selected DRAW element belongs will be stored. The draft to which the selected view belongs will also be stored.
- When an occurrence of a detail or symbol is selected, the detail workspace and contents will be stored.

Similar to checking the impact on model size in the previous examples, you can check the outcome after deleting elements with KEEP by selecting ERASE > PACK. Click on the YES button to pack the model. After erasing elements, you will notice that the numbers in the PACK window are different.

Another use of KEEP is merging passive models with the active model when working in a multi-model environment. This facility can be useful when you wish to combine all information in a single model upon working with models from different sources. To see how the merge option works, take the following steps.

1. Read the session you saved in Chapter 16 named *TROLLEY ASSEMBLY AUXVIEW2*.

2. Change the display to screen SC2 by clicking on the SC button, keying in *SC2*, and pressing <Enter>.

3. You have two methods of selecting the passive model to be merged. First, select KEEP > MERGE, and then choose an element of the passive model you wish to merge. All elements in the selected model will be highlighted. Click on the YES button to confirm the merge, and the passive model selected will be merged into the active model.

4. The second method involves selecting KEEP > MERGE, and pressing <Enter> (assuming the input information area is blank). From the list, select the passive model you wish to merge with the active model. All elements in the selected model will be highlighted. Click on the YES button to confirm the merge, and the passive model selected will be merged into the active model.

5. After merging, the passive model can be closed with FILE > CLOSE.

NOTE: When using the MERGE option, you cannot multi-select the required passive models. Selection and merging are carried out one model at a time.

The resulting model will have the same name as the original active model, unless you choose to save the resultant model with a new name.

Exercise 4: Using the IDENTIFY Function

IDENTIFY can be used to display, modify, and update the identifiers of all elements that have been used in a model. Upon viewing element identification in a model that you have cleaned, you will find that element numbering is not consecutive. At this point, you can rename the element identities to be consecutive by taking the following steps.

1. With the FILE > READ command, read a model that you have cleaned in any way.

2. Select IDENTIFY >RENAME | LIST | ELEMENT. Note that the element identifier numbering is not consecutive.

3. In order to make the element numbering consecutive, select IDENTIFY > RENUMBER | ELEMENT | AUTO ID. Click on the YES button to accept automatic renumbering.

4. Select IDENTIFY > RENAME | LIST | ELEMENT again. Note that the element identifier numbering is now consecutive.

Summary

This chapter covered the following topics: erasing elements from NO SHOW; deleting unwanted details and symbols; checking a model for errors and deleting errors; deleting from models everything other than selected elements; merging passive models with active models when working in a multi-model environment; and renumbering element identifiers.

Appendix A

Element Identifiers Used in CATIA

Appearing below are lists of DRAW element identifiers and SPACE element identifiers. All identifiers can be used for multi-selection. When using an identifier, use the asterisk (*) as a prefix (e.g., *PTD).

DRAW Element Identifiers

Element Identifier	Description
ANND	New dimensions and text
ARWD	Mark-up arrow
AXSD	Axis
CIRD	Circle
CND	Schematics connector
CRVD	All types of curves
DIM	All types of dimensions (old and new)
DIMN	New dimension
DITD	Ditto (occurrence of a detail)
ELLD	Ellipses
GDLN	Line grid
GDPT	Point grid
HYPD	Hyperbola

Element Identifier	Description
LND	Line
LST	Line string
OCND	Occurrences in schematics
PARD	Parabola
PTD	Point
SHAP	Shape (pattern hatching)
SPLD	Spline
STRD	Contour
SYMD	Occurrence of a symbol
TXTD	All types of text (old and new)
TXTN	New text
VU	View

SPACE Element Identifiers

Element Identifier	Description
AXS	Axis
CCV	Composite curve
CNP	Tubing connector
CRV	Curve
CST	Constraint
DIT	Ditto (occurrence of a detail)
FAC	Face
FSUR	Functional surface
LN	Line
NET	Network
OCP	Occurrence of piping
PIP	Pipe
PLN	Plane
POL	Polyhedral surface
PT	Point
SKI	Skin
SOE	Exact solid

SPACE Element Identifiers

Element Identifier	Description
SOL	All solids (exact and mock-up)
SOM	Mock-up solid
STR	Structure
SUR	Surface
TXT	All types of text (old and new)
VOL	Volume

> **NOTE:** *All of the above keywords can be used when combining multiple selection keywords. For further information, see Appendix B.*

Appendix B

Keywords Used for Multiple Selection

Appearing below are lists of keywords that can be used for multiple selection. When using a keyword, insert an asterisk (*) as a prefix. (For instance, to select all DRAW elements, you would key in *DRW.)

Single Interaction Keywords Requiring No Additional Information

Keyword	Description
DRW	All DRAW elements
SPC	All SPACE elements

All types of element identifiers (listed in Appendix A) can also be used as single interaction keywords. Examples are *LND* and *LN*.

Single Interaction Keywords Requiring Additional Information

Keyword	Description
COLx	Color, where x = 1 to 125
GRPx	Group, where x = 1 to 3

Keyword	Description
LAYx	Layer, where x = 0 to 254
LNTx	Line type, where x = 1 to 32
THKx	Line thickness, where x = 1 to 6

➥ **NOTE:** *In the previous table, the range of available values appears in the "Description" column.*

The keywords in the above table require a value to enable use as single interactive keywords. For instance, the *COL3 keyword would select all green elements, and the *LNT2 keyword would select all dotted lines.

Keywords Requiring Complementary Interaction

Keyword	Description
COL	Color
GRP	Group
LAY	Layer
LIP	Elements of a piping logical line
LIS	Elements of a schematic logical line
LNT	Line type
OPL	Elements lying on a given plane
SEQ	Sequentially linked monoparametrics
SET	Elements of a set
THK	Line thickness
TYP	Type (typical)
VU	Elements of a view

Examples of usage for keywords requiring further definition by a complementary interaction follow.

- Key in *COL and select any element. All elements of that color will be selected.

- Key in *LAY and select any element. All elements in that layer will be selected.

- Key in *OPL* and select any element in SPACE mode. All elements lying on the same plane as the selected element will be selected.

- Key in *TYP* and select an element. All elements of that type will be selected.

Multiple Interaction Keywords

Keyword	Description
ITRP	Elements totally inside a trap
OTRP	Elements totally outside a trap
PRF	Profile
SEL	Individual selection of several elements
TRP	Elements partially or totally inside a trap
TRP?	Elements partially or totally inside a trap, but excluding elements defined by a further keyword
XTRP	Elements partially or totally outside a trap
$PAC	Parent faces of a VOL or SKI type element

Keywords in the previous table must be further defined by complementary interactions. Examples follow.

- After keying in *TRP*, indicate a series of points to define the required trap. When the trap is complete, click on the YES button to complete the trap definition. Click on YES again to confirm the trap selection.

- After keying in *SEL*, you will have to select a set of elements to include in the selection. When the selection is complete, click on the YES button to end the selection; click on YES again to confirm the selection.

Combining Multiple Selection Keywords

Any keyword listed in Appendices A or B can be combined in order to select several groups of elements in a single operation. Examples of combining keywords are listed below.

- Select all LN (SPACE lines) type elements that are green.

- Select all PTD (DRAW points) type elements and all LND (DRAW line) type elements.
- Select all DRW (DRAW) elements except CRV (curve) type elements.

To combine the keywords, you can use the three types of separators shown below.

- + (plus symbol) — Join the keywords preceding and following the symbol.
- - (minus symbol) — Subtract the keyword following the symbol with the keyword preceding it.
- & (ampersand) — Intersects keywords preceding and following symbol.

Examples of combined multiple selections follow.

- *LN+*COL3 selects all green SPACE lines.
- *PTD+*LND selects all DRAW points and all DRAW lines.
- *DRW-*CRV selects all DRAW elements except the CRV elements.
- *LN&*COL2 selects all red SPACE lines.

You can combine up to 10 keywords in a character string. Examples of combined multiple selections follow.

- *CRV+*LND&LAY10-COL2 selects all curves and lines lying on layer 10 except those that are red.
- *SOL+*VOL&LAY8 selects all solids and volumes that are on layer 8.
- *SOL-*COL3 selects all solids except those that are green.
- *SOL-*COL3+*LN selects all solids except green ones, while also selecting all lines.
- *SPC-*SOL selects all SPACE elements except solids.
- *SOL+*VOL selects all solids and volumes.
- *PT+*LN+*CRV+*PLN selects all SPACE points, lines, curves, and planes.

Combining Multiple Selection Keywords

When using combined multiple selections, the following rules and general observations are pertinent.

- A maximum of ten keywords can be combined in a character string.

- A separator must not precede the first keyword.

- The string of keywords is analyzed from left to right.

- If an incorrect string of keywords is entered, the following message will be displayed: *IDENTIFER DOES NOT EXIST.*

- Combining keywords requiring graphical interaction after they have been entered is not possible. (An example is **TRP.*)

- The most commonly used combinations of keywords are available for use direct from the SELECT pull-down menu. For further information, see Appendix E.

Appendix C

CATIA General Commands

Appearing below is a list of the most popular general commands. Key in a forward slash before the general command. For example, to clean a model, key in /CLN, and follow the resulting prompts.

General commands are available regardless of the function you are using. Some of the commands can be used only in specific modes, such as SPACE 3D or 2D, or DRAW. Many general commands are also available in pull-down menus or via fixed menu buttons.

General Command Descriptions

Command	Description
ANACRVT	Analysis of element curvature parameters.
ANADEL	Delete an analysis applied to an element.
ANADRAFT	Check to determine whether an element can be correctly drafted.
ANND34	Upgrade V3R2 dimensions to new dimensions.
CLN	Clean model.
CLOSE	Close model.
COLCHG	Change color of an element.

Command	Description
COLCOP	Copy the color of an element.
COLSTD	Reset the color of an element to the color set in STANDARD.
COMMENT	Model comment display and edit (old style dialog).
COMMENT2	Model comment display and edit (motif style dialog).
CONFIG	Display code level information.
DIMCOORD	Add a box with point coordinates and leader to point (DRAW mode).
DOC	Search available documentation.
EXIT	Close the current session.
GRAB	Generate an image from the screen display using dialog window.
GRABP	Generate an image direct from screen display.
HELP	Start help mode window.
INFO	Display information about current model.
MDLALL	Apply ALL filter.
MDLFIL	Change current layer filter.
MONAXE	Change axis system (AXS button).
MONBRG	Buffer regenerate (BR button).
MONELE	Display element identifiers (ID button).
MONGS	Set graphic standards (ST button).
MONKEY	Define keyboard layouts (KEY button).
MONLAY	Change current layer (Lxxx button).
MONLST	Change between display modes, wireframe, hidden line, or shaded (Set current display mode button).
MONNP	Modify the pick/no pick attributes of elements (NP button).
MONNS	Modify the show/no show attributes of elements (NS button).
MONROT	Rotate the model.
MONS3D	Change between the SPACE 2D and 3D modes (2D/3D button).
MONSCR	Retrieve a stored screen (SC button).
MONSET	Change current set (SET = button).
MONSDR	Change between SPACE and DRAW modes (SP/DR button).
MONV2D	Change current DRAW view (VU button).
MONWSP	Change current workspace (WSP button).
MONWIN	Retrieve a stored window (WI button).

General Command Descriptions

Command	Description
MONZOM	Change image scale (ZM button).
NEW	Create a new model.
NEWS	Review the latest enhancements.
OPEN	Open a model.
PLOT	Plot the current model or models as viewed on the screen.
RECOVERY	Switch on recovery mode for working session.
REFRAME	Image reframe.
REMOTE	Remote access to mainframe files.
SAVE	Save model.
SAVEAS	Save model with a new name.
SAVESES	Save session.
SAVESESA	Save session with a new name.
SETUP	Modify graphics options, cursor type, fast transforms, and preselect highlighting.
SOLPATT	Create pattern of common features when creating solids.
SOLPAR	Generate basic geometric entities from solid primitives.
SOLSIZE	Display data size of a solid.
STCOLD	Set DRAW mode color standard.
STCOLS	Set SPACE mode color standard.
STRETCH	Modify 2D profile with stretching functions.
SWAP	Swap active model.

➙ **NOTE 1:** *Availability of some of the above commands is dependent on your CATIA setup.*

➙ **NOTE 2:** *When using general commands, you may not need to enter the entire command name. In many instances, you need only enter sufficient characters for CATIA to recognize the command (e.g., /STR for /STRETCH).*

Appendix D

Default Function Palette

The following illustration presents the default layout for the lighted program function keyboard (LPFK). These functions appear on the default LPFK palette.

Default LPFK palette.

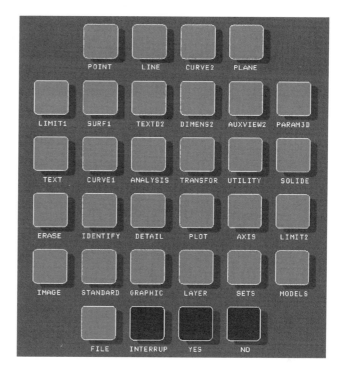

Appendix D: Default Function Palette

Up to five other palettes can be stored. See Chapter 3 for details on how to create and store alternative palettes. In this book, one alternative palette (MIX) was used, shown in the next illustration.

MIX palette layout.

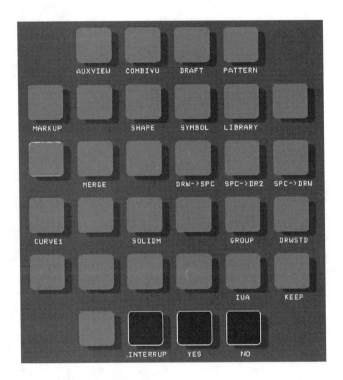

Appendix E

CATIA Pull-down Menus

The pull-down menu bar at the top of the CATIA screen provides access to a series of tools. Appearing below are lists of tools or options available in pull-down menus accompanied by usage descriptions. Keyboard access alternatives appear in parentheses.

File

The File pull-down menu contains tools for the management of models, files, and sessions.

File pull-down menu.

File Pull-down Menu Options/Tools

Option/Tool	Description
New	Create a new model or session (<Ctrl>+<N>).
Open	Open an existing model or session (<Ctrl>+<O>).
Open Database	Open a database (if available).
Close	Close a model.
Save	Save the current model with current name (<Ctrl>+<S>).
Save As	Save the current model with option to specify a new name.
Save All	Save all models in current multi-model environment or session.
Information	Display information about current model or session (<Ctrl>+<I>).
Save Session	Save session using the Open window.
Save Session As	Save session with a new name using the Open window.
File Manager	Copy, move, rename, or delete models.
Print	Plot the current model(s) window or screen as in PLOT function (<Ctrl>+<P>).
Exit	Exit CATIA (<Ctrl>+<Q>).

Select

The Select pull-down menu provides a list of all element types used, which can also be accessed from the keyboard. (Some options access additional lists, such as SPACE elements.) For a complete listing of the definitions for element abbreviations used in CATIA, see Appendix B.

Select pull-down menu.

View

The View pull-down menu contains tools for managing model viewing, and provides fast access to these tools which are also available in various other CATIA functions.

View pull-down menu.

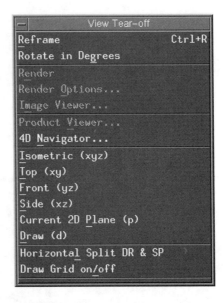

View Pull-down Menu Options/Tools

Option/Tool	Description
Reframe	Reframe the model to its extents (<Ctrl>+<r>).
Rotate	Rotate in degrees; available only in Version 4 R1.8.
Render	Access realistic rendering options and image viewer. Available only to licensees of CATIA Realistic Rendering.
Product Viewer	View the graphical version of an assembly imported from a database. Available only to licensees of CATIA Data Management Access.
4D Navigator	View by navigation through model.
Isometric (xyz)	View model on XYZ.
Top (xy)	View model from above on XY plane.
Front (yz)	View model from front on YZ plane.
Side (xz)	View model from side on XZ plane.
Current 2D Plane (p)	Move view on current 2D plane parallel to screen.
Draw (d)	Change to DRAW window.
Horizontal Split DR and SP	Split screen horizontally to incorporate SPACE and DRAW windows. Available only in Version 4 R1.8.

Option/Tool	Description
Draw Grid on/off	Enable creation of a point grid in DRAW mode. Available only in Version 4 R1.8.

Note: A list of windows stored in the model will also be displayed at the bottom of the View pull-down menu.

Filter

The Filter pull-down menu enables you to apply filters to a model.

Filter pull-down menu.

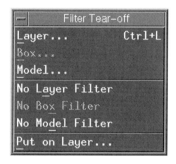

Filter Pull-down Menu Options/Tools

Option/Tool	Description
Layer	Apply a layer filter (<Ctrl>+<L>).
Box	Requires third party Box product.
Model	Apply a filter displaying only the current model.
No Layer Filter	Delete the previously applied filter to the current layer.
No Box Filter	Requires third party Box product.
No Model Filter	Delete the previously applied filter to the model.
Put on Layer	Transfer an element to a different layer. Available only in V 4 R1.8.

Options

The Options pull-down menu is used to modify palette positions. (See Chapter 1.)

Options Pull-down Menu Tools

Tool	Description
Full Screen Layout	View maximum possible work area size without palettes or current function displayed.
Palette Placement	Change placement of palettes and function menus.

Options pull-down menu.

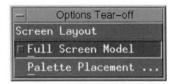

Tools

The Tools pull-down menu provides fast access to many commands that are also available from the keyboard and within other CATIA functions.

Tools Pull-down Menu Options

Option	Description
Change Color	Access color selection as in GRAPHIC function.
Copy from/to	Copy files from/to.
Reset to standard	Reset to standard colors.
Set SP Element Color	Set SPACE mode element colors.
Set DR Element Color	Set DRAW mode element colors.
Sketcher	Activate the Dynamic Sketcher (<Ctrl>+<K>).
Assembly	Create and manage multi-level assembly of parts.
Grid for Feature	Create or edit feature patterns (for use with part design V 4 R1.8).
Geometry	Create draw points, space points, draw lines, or space lines.
Visualization	Modify visualization of elements or refresh view.
Analyze	Analyze elements, clearances, and clashes.
Management	Write an object into a library, analyze space lines, set up configuration, mount files.
IUA Commands	Access IUA commands. (See Appendix F.)

Window

Option	Description
Conciliation	Activate Conciliation feature for use with concurrent engineering.
Engraving	Access Engraving/Embossing product if available.
Screen Grab	Enable TIFF images to be captured from the screen. (See Chapter 8.)
Insert/Edit Hyperlinks	Link external multimedia objects to CATIA elements.
Update	Update exact solids.

Tools pull-down menu.

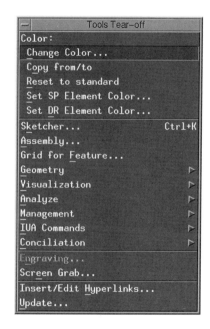

Window

The Window pull-down menu lists models contained in the current session and provides tools for managing overlaid models.

Window pull-down menu.

Window Pull-down Menu Options/Tools

Option/Tool	Description
Highlight Active Model	Highlight active model in red.
Show Active Only	Show active model only.
Show/Hide Models	Show/hide specific models.
Swap by Graphic Selection	Change a passive model to active status by selecting it.
Keep Current Layout	Change active model but maintain current screen layout if required.
Keep Current Axis	Keep current axis (or not) on swapping to passive.

Help

The Help pull-down menu provides access to various online help facilities.

Help pull-down menu.

Help Pull-down Menu Options/Tools

Option/Tool	Description
Getting Help	Activate online help.
Reference Doc	Activate only CATIA reference documentation in Frameviewer.
News in this Release	Display CATIA Solutions Product Enhancements Overview.
Mouse & Keyboard	Display a summary of mouse and keyboard information.
Computer Based Training	Access computer based online training information.

Appendix F

CATIA Utilities

Utilities can be accessed by double-clicking on the CatUtil icon in the CATIA_V4 window on the desktop, or by selecting the UTILITY function from the palette.

> **NOTE:** *To quickly swap back to the desktop, press and hold the <Alt> key and then press the <Tab> key, or select the Minimize button at the top right corner of the screen.*

Appendix F: CATIA Utilities

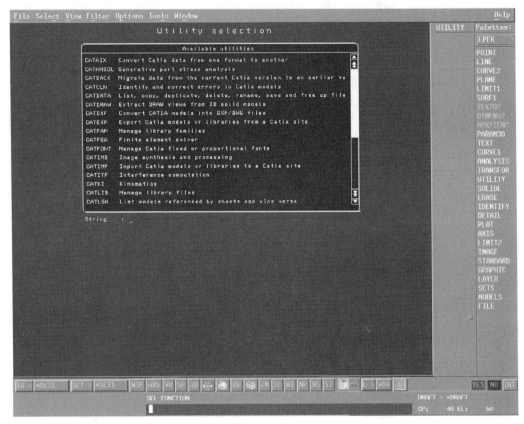

Utility selection screen (top).

CATIA Utilities

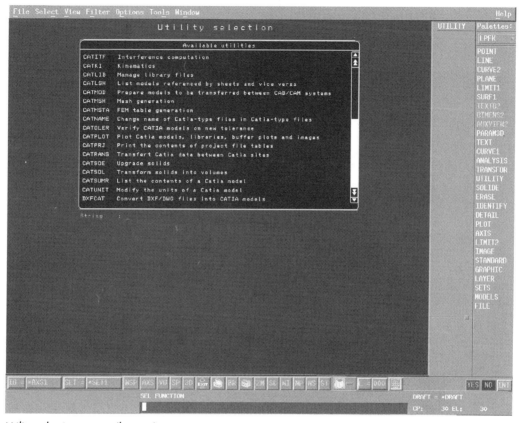

Utility selection screen (bottom).

Appearing below is a list of the available utilities and a brief description of their use.

CATAIX	Convert CATIA data produced on IBM S/370 computers in EBCDIC format to ASCII format for IBM RISC System/6000 computers.
CATANSOL	*For use with ANSOLID function.
CATBACK	Migrate CATIA data from one version/release to an earlier version/release.
CATCLN	Check for errors in a specific model. If errors are found, correction measures are proposed and can be applied.
CATDATA	Perform the following functions relating to CATIA files stored on tape or disc: print, copy tape to disc (or vice versa), duplicate, rename, delete, and save.

CATDRAW	Update DRAW views similar to AUXVIEW2 I UPDATE.
CATDXF	Converts CATIA models (not including solids, volumes, skins, and some other complex entities) to DXF or DWG format for use with other CAD systems.
CATEXP	Extracts CATIA data in preparation for transfer to another site to avoid conflicts with project file at different sites.
CATFAM	Creation and management of CATIA Library families.
CATFEA	*For use with CATIA Finite Element Solver.
CATFONT	Create special character fonts.
CATIMD	*Generate high quality images as in SHADES function in batch mode.
CATIMP	Imports previously exported data from another site. (This utility is used to load the exercise models from the CD.)
CATITF	Compute interferences between CATIA elements.
CATKI	*Used in conjunction with CATIA Kinematics.
CATLIB	Provides additional facilities in CATFAM for the management of CATIA libraries.
CATLSH	Prints information from SHEET files (where filed plots are stored). For instance, models list is referenced by each sheet and the list of sheets in each model.
CATMOD	Perform operations on CATIA models in order to prepare them for transfer to other CAD systems.
CATMESH	*Used in conjunction with CATIA finite element analysis products.
CATMSTA	*Used in conjunction with CATIA finite element analysis products.
CATNAME	Change CATIA file names.
CATOLER	Analyze 3D CATIA elements, and then modify elements to make them compatible with newly defined tolerances. (Used in conjunction with CATRANS.)
CATPLOT	Works like /PLOT command. Enables plotting of SHEET, PICTURE, and LIBRARY files as well as buffer plots.
CATPRJ	Print the contents of tables in the project file.
CATRANS	For transfer of CATIA models between previously synchronized sites.
CATSOE	Analyze and upgrade CATIA Version 3/Release 2 solids in a model (works as /M SOLM34).
CATSOL	Create a complex volume from a solid.
CATSUMR	List information about one or more CATIA models including date and time of creation, execution instructions, messages, associated project file, comments, and model standards.
CATUNIT	Modify units of some or all contents of a CATIA model.
DXFCAT	Coverts DXF or DWG files to a CATIA model.

* Utilities used in applications not covered by INSIDE CATIA.

In addition, another set of utility tools is available by selecting the IUA (interactive user access) commands function as follows: Select IUA > EXECUTE, and then select the required item from the list of about 300. Alternatively, any of the items can be transparently selected (i.e., while working within a CATIA function) by keying in */M iua command name from list* (e.g., */M AND2PT* to perform relative analysis between two points).

Appendix G

Engineering Symbols

When annotating drawings, having a range of engineering symbols available is useful. Unfortunately, engineering symbols are not available on many UNIX keyboards. However, you can obtain such symbols by holding down <Ctrl>+<Alt> followed by another key.

Key Sequences for Available Engineering Symbols

Key in <Ctrl>+<Alt> followed by	q	w	e	r	t	y	u	i
Resultant symbol	∉	∅	°	Ω	≤	≥	±	μ

Index

Symbols

/CLN command 351
   usage 354–355

Numerics

2D elements
   creating 15–23
   setting up workstation for 15
2D in SPACE mode 202
3D in SPACE mode 202

A

ANALYSIS function
   analyzing solids 251–253
ANALYSIS function, 2D SPACE mode 216
ANALYSIS function, DRAW mode
   alphanumeric window 51
   options 49
   RELATIVE option 53
ANALYSIS function, SPACE mode
   options 215
   usage 216–220
assemblies
   adding models 296
   modifying 325
assembly analysis 260
autodimensioning 341
AUXVIEW function
   creating isometric views 112–115
   creating orthographic views 94–111
   options 90
   usage 310–314
   usage with DRAFT 122
AUXVIEW function, DRAW mode
   usage 166–167
AUXVIEW2 function

Action window 304
Defaults dialog 307
description 303
Objects window 304
usage 314–325
view options 305
AXIS function
   options 93
AXIS function, DRAW mode
   usage 116–120

B

blink
   element display 40
boundary representation (B-REP) 221
BREAKOUT function 298, 331
B-REP (boundary representation) 221

C

CATIA News panel 5
CATIA session, beginning 5
chamfers
   creating 35
circle
   creating by defining centers with coordinate
      dimensions 43
cold start 5
color 39
COMBIVU function
   creating isometric view 112–115
   creating orthographic views 94–111
   options 91
computational sag 259
computer hardware
   basic 1
   minimum 1

mouse, three-button 1
cone
  creating 229
constructive solid geometry (CSG) 221
creating geometry
  SPACE mode 204–208
CSG (constructive solid geometry) 221
cuboid
  creating 227–228
CURVE2 function, DRAW mode 14
  creating circles 20
  options 15
CURVE2 function, SPACE MODE
  options in 3D 194
cylinder
  creating 230

D

Defaults dialog 307
deleting text 165
DETAIL function, DRAW mode
  options 62
DETAIL function, SPACE mode
  options 265
detail workspace
  transferring elements to 77
dials 2
  using for model manipulation 16
DIMENS2 function
  expanded Management dialog 137–141
  expanded Management dialog, buttons described 138–141
  parameter options 139
DIMENS2 function, CREATE options 134
DIMENS2 function, MODIFY options 136
dimensions
  parametric 343
discretization 222
dittos
  creating 78, 78–80
DRAFT function
  options 92
  usage with AUXVIEW 122
DRAFT function, DRAW mode
  usage 166
drafting 231
drafts
  changing 122
DRAW mode 8

DRAW views
  extracting with autodimensioning 341–343
drawing sheet blank
  creating 146–148
  creating symbol of 156–158
  inserting detail in master workspace 154–155, 170
DRW->SPC function, SPACE mode 278
DRWSTD function
  description 144
  Description dialog 144
  Numerical Display dialog 145
  options available to average users 144
  Symbol dialog 146
Dynamic Sketcher 347

E

element management functions 57
elements
  applying transformations 69
  color 39
  copying between models 298–??
  graphical representation, blinking highlights 40
  graphical representation, checking 40
  graphical standard, setting 10
  grouping 73
  line types 37
  management functions 57
  point types 39
  setting graphical representation 10
  thickness 39
  transferring to detail workspace 77
  transferring to layer 82–84
elements, transfer
  DRAW to SPACE 280–282
  SPACE to DRAW 285
engineering symbols 190
ERASE function
  housekeeping 352–354
ERASE function, DRAW mode
  example 31–32
  options 26
ERASE function, SPACE mode
  options 197
erasing
  details or symbols 353–354
exact revolution solid 244
exact solid prism 238

Index

exact solids 221
expanded Management dialog
   DIMENS2 function 137–141
   TEXTD2 function 129–131

F

FILE function 8
File pull-down menu 8, 292
fillet radius
   creating 232
filters
   usage with layers 82
finishing details 106–107
front view
   creating, orthographic views 94–99

G

geometry, transfer
   DRAW to SPACE 280–282
   SPACE to DRAW 285
GRAPHIC function, DRAW mode
   checking graphical attributes 40
   graphical representation, modifying 37–40
   options 28
GRAPHIC function, SPACE mode
   options 199
   usage 208
graphical representation
   dot-dashed lines 37
   line types 37, 38
   point types 39
   setting 10
   thickness 39
GROUP in DRAW mode
   options 60

H

housekeeping
   ERASE function 352–354

I

IDENTIFY function 350
   usage 357
IMAGE function in DRAW mode
   creating screens 85
IMAGE function, SPACE mode 274

interference analysis 259

K

KEEP function
   options 350
   usage 356–357
KEY X,Y method
   entering coordinate positions 17
keyboard cursor keys
   using for model manipulation 16

L

LAYER function, DRAW mode
   options 66
   transferring elements to layer 82
layers
   defined 82
   filter usage 82
Leader Management dialog 163
leaders
   creating 162–164
LIBRARY function, DRAW mode
   common uses 80
   options 64
lighted program function keyboard (LPFK) 3
LIMIT1 function
   using to relimit lines 31
LIMIT1 function, DRAW mode
   creating chamfers 36
   options 27
   relimiting corners 44
LIMIT1 function, SPACE mode
   options 198
LINE function, DRAW mode 13
   creating lines 18
   options 13
LINE function, SPACE mode
   options in 2D 192
   options in 3D 191
line thickness 39
line types 37, 38
lines
   creating 18–19
log-on procedure 3–5
LPFK (lighted program function keyboard) 3

M

Management dialog
  TEXTD2 function 128
MARKUP function, DRAW mode
  creating lines for curves 45
  options 29
MERGE function, SPACE mode
  options 267
mock-up solids 221
Model manipulation 16
models
  active 293
  breakout 298–299
  copying elements between 298
  creating blank 22
  creating new 32
  duplicate 295
  enlarging portion 85–87
  move 294
  passive 293
  rotate 294
  saving 19
  work areas 300–301
MODELS function, SPACE mode
  options 291
mouse
  four-button 2
multi-model environment
  creating 293–297
multi-select process 77
multi-selection
  options 119

N

notes
  creating 164–165

O

orthographic views
  creating, front view 94–99
  creating, plan view 108–111
  creating, side view 99–105

P

palette
  placement
    compact 7
    floating 7
    none 7
palettes 7
  creating 54
  creating a blank 54
  creating additional 45
  default 7
  Palette Creation screen 54
  placement 7
    permanent 7
  switching between 47
PARAM3D function, SPACE mode
  options 335
parameter options
  DIMENS2 function 139
parameterization
  autodimensioning 341–343
  creating profile 337–340
  positioning features 343–347
parameterized profile
  creating 337–340
parameters
  solids 237–238
parametric dimensions
  positioning features 343–347
PATTERN function
  cross hatching 168
  options 142
  usage 168
plan view
  creating, orthographic views 108–111
PLANE function, SPACE mode
  options 200
planes
  defining 210–211
PLOT function
  multiple window, file mode 183
PLOT function, DRAW mode
  full-size quick plot 176, 182
  multiple window quick plot 182–183
  multiple window, file mode 183–185
  multiple window, quick mode 182–183
  options 174
POINT function, DRAW mode
  creating points 17
  options 12
POINT function, SPACE mode
  options 190
  usage 208

Index

point types 39
points
  creating 17

R

relative analysis 53

S

saving model 19
saving sessions 296
screen
  split 281
screen capture 186–188
Screen Definition window 275
screens, capturing 186–188
sessions
  saving 296
sets
  duplicating 80–81
SETS function, DRAW mode
  function menu 65
SHAPE function 54–55
SHAPE function, DRAW mode
  options 50
side view
  creating, orthographic views 99–105
SOLIDE function
  ANALYSIS option, analyzing solids 253–260
  options 225
SOLIDM function
  options 223
solids
  breaking out within view 331–332
  creating a cone 229
  creating a cuboid 227–228
  creating a cylinder 230
  creating a fillet radius 232
  creating a nut 246–248
  creating a shell 235
  creating a sphere 231
  creating DRAW views from 287–289
  creating from DRAW geometry 282–285
  drafting 231
  duplicating 268, 269
  exact 221
  interference analysis 259
  mock-up 221
  modifying parameters 237–238

modifying visualization 329, ??–330
position analysis 257
positioning 268, 269
representations 221
solids analysis
  analyzing position of several 257–260
  using ANALYSIS function 251–253
  using ANALYSIS option in SOLIDE function 253–260
SP/DR toggle button 8
SPACE mode 8
  creating geometry 204–208
  creating text in 313–314
  setup 202
SPC->DR2 function, DRAW mode 280
SPC->DRW function, DRAW mode 279
sphere
  creating 231
SYMBOL function
  creating symbol of drawing sheet blank 156–158
symbols, engineering 390,
  viewing available 145
symmetry
  performing 72, 73–75

T

tear-off menu 8
  closing 9
text
  creating in SPACE mode 313–314
TEXT function, SPACE mode
  options 309
Text Management dialog 162
TEXTD2 function
  CREATE options 126
  expanded Management dialog 129–131
  expanded Management dialog, buttons described 129–131
  Management dialog 128
  MODIFY options 127
  orientation options 132–133
TEXTD2 Orientation Options dialog 132
title block
  annotating 150–154
  creating 148–150
toggle
  3D/2D 202
  SP/DR 8

TRANSFOR function, DRAW mode
  options 59
TRANSFOR function, SPACE mode
  options 263
transformation
  alternate methods 75–76
Transformation Apply window 71
transformations
  solid views 233–234
translation
  define 294

V

views
  auxiliary 320
  breaking out solids 331–332
  clipping 330
  copying 323
  creating with DRAW elements 316–325
  creating without DRAW elements 310–314
  Defaults dialog 307
  detail 319
  dimensioning 158–162
  filtering solids 326–329
  isometric 94, 312, 322, 329

modifying 324, 326–329, 326–333
modifying solids visualization 329–330
orthographic 94, 311, 314
plan 318
primary 316
principal 317
section 318
side elevation 318
solids, transformation 233–234
translating around screen 108
UPDATE option 325
updating 325
views, DRAW
  creating from solids 287–289

W

wireframe models
  creating 210–212
work area 6
work areas
  creating 300–301

Y

YOUR icon 4

CATIA Model Files on CD

Appearing below are descriptions and sizes of files recorded on the companion CD. Refer to "Installing Models from Companion CD-ROM" in Introduction for installation instructions.

File name	Size	Description
IC02EX1	2K	Target box with corner points model.
IC02EX2	2K	Start box with corner points model.
IC02EX2	2K	Target box with corner points, lines, and circles model.
IC02EX3	2K	Target SIMPLE EQUAL ANGLE WITH CORNER POINTS model.
IC02EX4	3K	Target SIMPLE I BEAM WITH CORNER POINTS model.
IC03EX1	2K	Target BOX CREATED WITHOUT POINTS model.
IC03EX3	2K	Target SIMPLE EQUAL ANGLE WITH CORNER RADIUS model.
IC03EX4	2K	Target SIMPLE I BEAM WITH CORNER RADII model.
IC03EX5	2K	Start SIMPLE I BEAM WITH CORNER RADII model.
IC03EX5	2K	Target SIMPLE I BEAM WITH CENTRE LINES model.
IC03EX6	3K	Target REAL I BEAM (UNIVERSAL BEAM) model.
IC03EX7	3K	Target SIDE PLATE GEOMETRY model.
IC04EX1	3K	Start I BEAM FOR ANALYSIS model.
IC04EX2	3K	Start I BEAM FOR ANALYSIS model.
IC04EX3	3K	Start I BEAM FOR SHAPE CREATION model.
IC05EX1	3K	Start I BEAM GEOMETRY model.
IC05EX2	2K	Start ONE QUARTER I BEAM model.
IC05EX3	2K	Start ONE QUARTER I BEAM model.
IC05EX4	3K	Start SIDE PLATE GEOMETRY model.

File name	Size	Description
IC05EX4	4K	Target SIDE PLATE GEOMETRY WITH DITTOS model.
IC05EX5	4K	Start SIDE PLATE GEOMETRY WITH DITTOS model.
IC05EX5	5K	Target SIDE PLATE DUPLICATE model.
IC05EX6	5K	Start SIDE PLATE DUPLICATE model.
IC05EX6	5K	Target SIDE PLATE DUPLICATE model.
IC05EX7	5K	Start SIDE PLATE DUPLICATE model.
IC05EX7	5K	Target 3 WINDOW SCREEN model.
IC06EX1	15K	Target 01 FRONT VIEW OF ANGLE model.
IC06EX1	16K	Target 02 FRONT AND SIDE VIEW OF ANGLE model.
IC06EX1	17K	Target 03 FRONT, SIDE AND PLAN VIEWS OF ANGLE model.
IC06EX1	18K	Target 04 FRONT, SIDE, PLAN AND ISO VIEWS OF ANGLE model.
IC06EX1	23K	Target 05 FRONT, SIDE, PLAN AND ISO VIEWS OF ANGLE (RAD) model.
IC06EX2	2K	Start SIMPLE EQUAL ANGLE WITH CORNER RADIUS model.
IC06EX2	2K	Target DUPLICATE ANGLE AFTER MOVE model.
IC06EX3	2K	Start DUPLICATE ANGLE AFTER MOVE model.
IC06EX4	2K	Start FRONT, SIDE, PLAN AND ISO VIEWS OF ANGLE model.
IC06EX4	385K	Target CREATING A NEW DRAFT model.
IC07EX1	3K	Start SIDE PLATE model.
IC07EX1	7K	Target 01 TITLE BLOCK WITH TEXT POINTS model.
IC07EX1	23K	Target 02 SIDE PLATE AND DRAWING SHEET model.
IC07EX1	31K	Target 03 USE OF SYMBOL INSTEAD OF DETAIL model.
IC07EX1	41K	Target 04 SIDE PLATE WITH DIMENSIONS model.
IC07EX1	43K	Target 05 SIDE PLATE MANUF DRWG WITH DIMENSIONS AND NOTES model.
IC07EX2	57K	Target 01 WHEEL DRAFT CREATED WITH NEW PLOT SHEET model.
IC07EX2	71K	Target 02 SIDE PLATE AND WHEEL MANUF DRWG COMPLETE model.

File name	Size	Description
IC09EX1	15K	Target 3D I BEAM GEOMETRY model.
IC09EX2	16K	Target 3D SIDE PLATE GEOMETRY model.
IC09EX3	15K	Target 3D WHEEL GEOMETRY model.
IC09EX4	15K	Start 3D I BEAM GEOMETRY model.
IC09EX4	20K	Target 3D I BEAM WIREFRAME model.
IC11EX1	9K	Target 01 CUBOID SOLID model.
IC11EX1	15K	Target 02 CUBOID AND CONE model.
IC11EX1	20K	Target 03 CUBOID, CONE, AND CYLINDER model.
IC11EX1	28K	Target 04 CUBOID, CONE, CYLINDER, AND SPHERE model.
IC11EX1	34K	Target 05 CUBOID, CONE, CYLINDER, SPHERE, AND DRAFT model.
IC11EX1	39K	Target 06 COMBINED SOLID AFTER UNION model.
IC11EX1	80K	Target 07 SOLID WITH FILLETS model.
IC11EX1	168K	Target 08 SHELL SOLID model.
IC11EX1	198K	Target 09 SHELL SOLID CUT IN HALF model.
IC11EX1	299K	Target 10 SHELL SOLID CUT IN HALF WITH 4MM THICKNESS model.
IC11EX1	198K	Target 11 HALF SHELL COMPLETE WITH MODIFICATIONS model.
IC11EX2	15K	Start 3D I BEAM GEOMETRY model.
IC11EX2	43K	Target BEAM TROLLEY SOLID I BEAM SOLID ON LAYER 10 model.
IC11EX3	16K	Start 3D SIDE PLATE GEOMETRY model.
IC11EX3	77K	Target 1 SOLID SIDE PLATE WITH 4 HOLES model.
IC11EX3	132K	Target 2 SOLID SIDE PLATE COMPLETE ON LAYER 10 model.
IC11EX4	15K	Start 3D WHEEL GEOMETRY model.
IC11EX4	50K	Target 01 SOLID WHEEL FIRST STAGE model.
IC11EX4	115K	Target 02 SOLID WHEEL WITH BASIC WEBS model.
IC11EX4	352K	Target 03 SOLID WHEEL WITH BASIC WEBS + FILLETS model.
IC11EX4	409K	Target 04 SOLID WHEEL COMPLETE LAYER 10 model.

File name	Size	Description
IC11EX5	18K	Target 01 SOLID NUT BASIC HEXAGON model.
IC11EX5	42K	Target 02 SOLID NUT COMPLETE ON LAYER 10 model.
IC11EX5	23K	Target 03 SOLID BEARING LAYER 10 model.
IC11EX5	20K	Target 04 SOLID BUSH LAYER 10 model.
IC11EX5	20K	Target 05 SOLID SPINDLE LAYER 10 model.
IC11EX5	15K	Target 06 SOLID STUD LAYER 10 model.
IC11EX5	12K	Target 07 SOLID WASHER LAYER 10 model.
IC12EX1	43K	Start 01 BEAM TROLLEY SOLID I BEAM ON LAYER 10 model.
IC12EX2	43K	Start 01 I BEAM SOLID ON LAYER 10 model.
IC12EX2	431K	Start 02 SOLID WHEEL FOR ANALYSIS model.
IC12EX3	431K	Start 01 SOLID WHEEL FOR ANALYSIS model.
IC12EX3	1247K	Start 02 SOLID WHEELS FOR ANALYSIS model.
IC13EX1	132K	Start SOLID SIDE PLATE COMPLETE ON LAYER 10 model.
IC13EX1	250K	Target SIDE PLATES CORRECTLY POSITIONED model.
IC13EX2	409K	Start SOLID WHEEL COMPLETE LAYER 10 model.
IC13EX2	412K	Target BEAM TROLLEY 4 WHEELS COMPLETE model.
IC13EX3	412K	Start BEAM TROLLEY 4 WHEELS COMPLETE model.
IC13EX3	80K	Target 01BEAM TROLLEY WHEEL SPINDLE ASSEMBLY model.
IC13EX3	490K	Target 024 WHEELS + SPLINDLE ASSY AFTER MERGE model.
IC13EX3	493K	Target 034 WHEELS + SPLINDLES COMPLETE model.
IC13EX4	493K	Start 4 WHEELS + SPLINDLES COMPLETE model.
IC13EX4	493K	Target 4 WINDOW SCREEN OF4 WHEELS + SPLINDLES COMPLETE model.
IC14EX1	3K	Start I BEAM GEOMETRY model.
IC14EX1	40K	Target I BEAM GEOMETRY AND SOLID model.
IC14EX2	43K	Start SIDE PLATE MANUF DRWG WITH DIMENSIONS AND NOTES model.
IC14EX2	92K	Target SIDE PLATE SOLID model.

File name	Size	Description
IC14EX3	40K	Start I BEAM GEOMETRY AND SOLID model.
IC14EX3	6K	Target I BEAM GEOMETRY model.
IC14EX4	409K	Start SOLID WHEEL COMPLETE LAYER 10 model.
IC14EX4	449K	Target SOLID WHEEL AND DRAW VIEWS model.
IC15EX1	1K	Start 01 EMPTY MODEL WITH CURRENT LAYER 10 model.
IC15EX1	80K	Start 02 BEAM TROLLEY WHEEL SPINDLE ASSEMBLY model.
IC15EX1	412K	Start 03 BEAM TROLLEY 4 WHEELS COMPLETE model.
IC15EX1	250K	Start 04 BEAM TROLLEY SIDE PLATES COMPLETE model.
IC15EX1	117K	Start 05 BEAM TROLLEY STUD ASSEMBLY model.
IC15EX1	43K	Start 06 BEAM TROLLEY SOLID I BEAM model.
IC15EX2	397K	Target WHEEL COPIED FROM 4 WHEEL MODEL model.
IC15EX3	43K	Start BEAM TROLLEY SOLID I BEAM model.
IC15EX3	43K	Target USING BREAKOUT model.
IC16EX1	412K	Start BEAM TROLLEY 4 WHEELS COMPLETE model.
IC16EX1	434K	Target 01 CREATING VIEWS OF 4 WHEELS MODEL model.
IC16EX1	432K	Target 02 CREATING VIEWS OF 4 WHEELS MODEL WITH TEXT model.
IC16EX2	447K	Target 01 VIEWS OF BEAM TROLLEY ASSEMBLY model.
IC17EX1	43K	Start I BEAM SOLID WITH FILTER ALL APPLIED model.
IC17EX1	67K	Target 01 PARAMETERIZED I BEAM model.
IC17EX1	67K	Target 02 PARAMETERIZED I BEAM MODIFIED model.
IC17EX2	67K	Start MODIFIED I BEAM model.
IC17EX2	95K	Target 01 MODIFIED I BEAM + VIEW WITH AUTO DIMENSIONS model.
IC17EX2	95K	Target 02 MODIFIED I BEAM + ORIGINAL DIMENSIONS RESTORED model.
IC17EX3	132K	Start SIDE PLATE SOLID WITH FILTER ALL APPLIED model.
IC17EX3	156K	Target 01 PARAMETIZED SIDE PLATE model.

File name	Size	Description
IC17EX3	195K	Target 02 PARAMERTIZED SIDE PLATE MODIFIED HOLE POSITION model.

NOTE: The CD-ROM accompanying this book does not contain CATIA software

OnWord Press Distribution

OnWord Press books are available worldwide from OnWord Press and your local bookseller. For order information, terms, or listings of local booksellers carrying OnWord Press books, call toll-free 1-800-4-ONWORD (1-800-466-9673) or 505-474-5130; fax 505-474-5030; write to OnWord Press, 2530 Camino Entrada, Santa Fe, New Mexico 87505-4835, USA, or e-mail orders@ hmp.com. OnWord Press is a division of High Mountain Press.

Comments and Corrections

Your comments can help us make better products. If you find an error, or have a comment or a query for the authors, please contact us at the address below, send e-mail to cleyba@hmp.com or call us at 1-800-466-9673.

OnWord Press, 2530 Camino Entrada, Santa Fe, NM 87505-4835 USA

On the Internet: http://www.hmp.com